AF718992

THE DOUBLE SHELL SOLAR HOUSE

Community Builders' Experience with the Buffering Air Envelope

Three Houses by Hank Huber

Edited Discussion from Three Envelope Technical Conferences

Published by Community Builders, Canterbury, NH 03224

Community Builders
Quality Construction, Energy Conserving/Solar Design

(C) 1980 by Community Builders. All rights reserved. No part of this book may be reproduced in any form without the written permission of the publisher.

ISBN 0-9604422-1-9

TH
7414
.D68

Some of our work was made possible by a grant (#DE-FG1-79R110033) from the US Department of Energy's Appropriate Technology Small Grants Program. Any opinions, findings, conclusions, or recommendations expressed herein are those of the authors and do not necessarily reflect the views of DOE.

WITHDRAWN

CONTENTS:

GENERAL EDITOR: Don Booth

PERFORMANCE AND SIMULATIONS CHAPTERS: Doug Clayton

DISCUSSIONS AND GRAPHS: Bob Henninge

DRAWINGS: Donald A. King; Deanne Bednar and Barbara Putnam

COVER: Marianne Bogen

Parts of this book include information shared at three conferences of envelope designers and builders held during 1979 and 1980. We have quoted or paraphrased many contributions, and identified the speakers where possible. We are grateful to the following, and others:

March 1980 Conference Panel

GB- Galen Beale, Loudon, NH
DB- Don Booth, Community Builders
BB- Bill Brodhead, designer/builder, Buffalo Homes, Riegelsville, PA
DC- Doug Clayton, Community Builders
DD- Dennis Davey, designer/builder, Willington, CT
JF- John Flippen, designer/builder, Cartersville, VA
BH- Bob Henninge, Community Builders
HH- Hank Huber, designer/builder, New Ipswich, NH
WM- William Mead, architect, WM Design Group, Center Harbor, NH
NS- Norman Saunders, P E, Weston, MA

Other participants quoted:

JA- Joshua Arnow, New York, NY
AC- Al Converse, Dartmouth College, Hanover, NH
HG- Homer Ghaffari, Brookhaven National Laboratory, Upton, NY
DH- Doug Holmes, Lexington, MA
MK- Mark Kelley, Acorn Structures, Concord, MA
RM- Robert Mastin, Ekose'a, Middletown, RI
PP- Paul Peters, designer/builder, Sellersville, PA
RS- Robert Smith, P E , Newton Highlands, MA
GT- Gordon Tully, architect, Massdesign, Cambridge, MA

Plus:

JB- Jim Berk, Solar Clime Designs, Stanford, CA, editor of "Convection Loops," who has been very effective in providing a forum for envelope designers and whose "Working Papers" are defining design principles. Although Jim was not present, he was often quoted at the conferences.

1. Introduction

by Don Booth

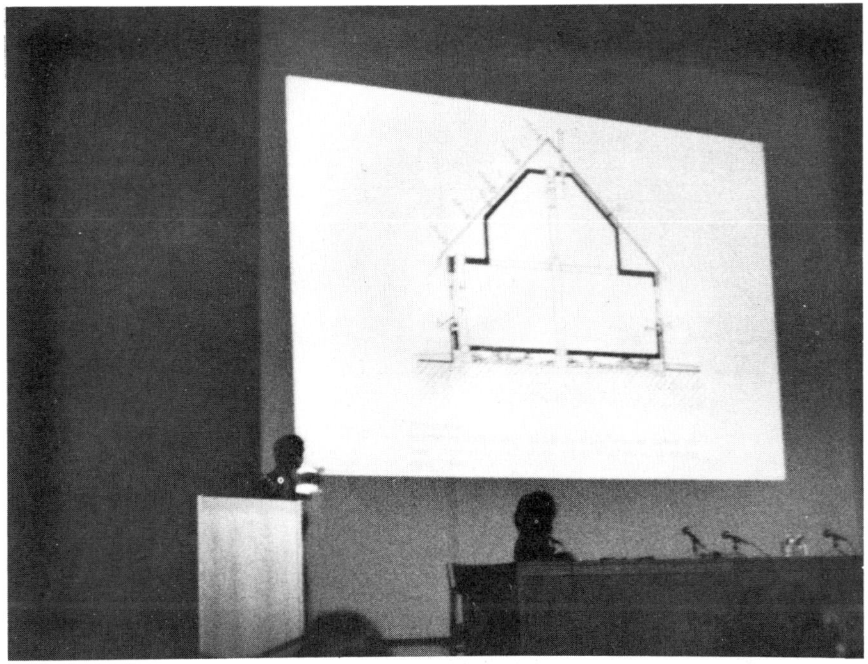

This photo happens to show the moment of my introduction to the buffering air envelope in 1978 at a Stockholm solar heating conference.

I heard Johannes Gunnarshaug describe a Norwegian system which used a "building within a building" to form an airspace between the inner and outer shells so that the inner building could stand in an artificially tempered climate. Gunnarshaug reasoned that even the faint sunshine of Norwegian winters could produce enough heat to make a buffering air envelope warmer than the outdoor air, and that this warmer climate around the occupied inner building would importantly reduce the heating load on the inner space.

During that trip I also saw Bengt Warne's "Nature House" near Stockholm, shown at right. Warne put a transparent protective shell around a carefully controlled self-sufficient living space. Plants grow food and purify air, sunshine is used by plants and for heat, a compost toilet returns material to the earth. A woodstove uses natural convection airflow to distribute supplementary heat.

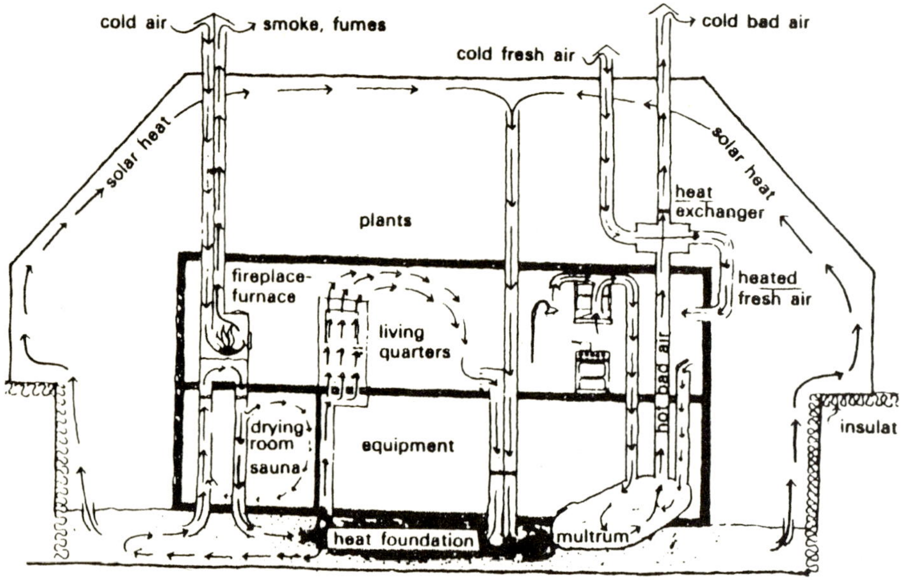

During the summer of 1978 I was impressed with the Co-Evolution Quarterly description of Tom Smith's Tahoe house, and when I was asked to coordinate the planning of a passively heated school building I soon turned to double shell. The Tahoe house demonstrated Lee Porter Butler's discovery that magically silent gravity airflows could carry greenhouse solar heat around an inner building, even storing some of that heat in the crawlspace.

We had built several sun-heated homes, using direct gain, water wall, and internal convective loop systems, and knew there are many good and simple ways to use sunshine effectively. Several factors led us to an air envelope for the Sant Bani Ashram School (drawings are on page 10.5): a school can't use much heat during the daytime, because of large internal heat gains from people and lighting; a greenhouse buffering zone is cheerful, soothing, inspiring, and provides exciting opportunities for student participation; the natural processes are integral, obvious, teaching. (The building was not constructed because the school had an unexpected opportunity to buy an existing nearby building at a favorable price.)

Available explanations of the Tahoe performance did not seem useful. The concepts seemed ideally promising, but we felt we had to undertake our own studies to answer our many questions. How much air could flow around the loop? How much heat could be transferred to earth or bedrock? How much auxiliary heat would the inner space need?

Doug Clayton arrived from two years at Farallones Institute, and helped study the envelope's possibilities with TEANET and ASHRAE. We also won a DOE Appropriate Technology grant for design and study of a double shell house - the Burns house, described on the following pages.

I think a large part of the present controversy about the air-envelope system is due to the extravagant and misleading descriptions that are still being written about its performance. Experienced solar designers have good reason to be upset about the claims. Some statements are not extravagant, when we consider facts like the Burns house's winter use of a tenth of a cord of firewood for supplemental heat and the still unproven conviction of some designers that we can do much better than that - but this is not yet the universally ideal system. I've seen many envelope houses that use much more fuel, two with envelopes that can't work at all, and many that I think will have high maintenance costs. The principles are simple but must be awarely applied.

The physics of the transportation of heat by natural convection within the double shell became much more clear through measuring and graphing temperatures, and through the pooling of experience by designers, builders and owners that took place in a series of technical conferences we arranged in 1979 and 1980.

Physics is replacing magic.

2. Burns House

Warren and Mary Burns wanted an energy-efficient house for retirement living--a low-profile, one-story house that would give them what they'd been reading about in Mother Earth News.

One thing we thought when we looked at the sloping site for this house was: would water run under the house and take the heat away as fast as we could push it into the ground storage? We thought for awhile about digging

a ten foot deep trench, six or eight feet north of the building on the hillside, and draining all the water away, but when we considered how difficult it would be to do that we decided instead just to put in a normal footing drain around the actual structure. We're always very careful with drainage. Many people don't even recognize that the holes in the perforated pipe go at the bottom. When these pipes are carefully and properly installed, they're self-cleaning--there's a little trough at the bottom between the two rows of holes that will take water away. We also protect the drainage with clean crushed stone, so that water coming down through the earth to the drain is going to be cleaned on its way.

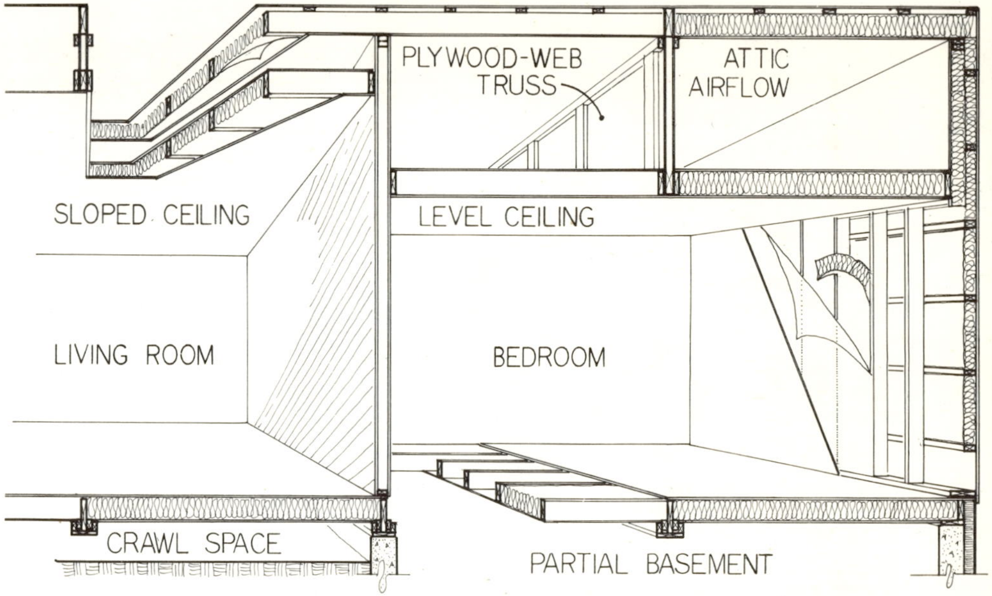

We felt it would be helpful to increase the useful earth mass by putting some of the earth outside the building into contact with the foundation wall, so we put our insulation three feet out from the wall on the north. That required tamping carefully on both sides of the insulation--the trapezoids here (2.2 at upper right) are simply spacers that were used to help us keep the polystyrene straight. I have been aware that the blue extruded Styrofoam is recommended by its manufacturer as the only insulation appropriate for use below grade, but I don't like its high cost so I've been hoping that the cheaper beadboard (at less than half the cost) will stay dry enough for it to be a better choice on a cost basis. I've depended on a good surface-water drainage through sand or gravel to an effective footing drain. Recently I've learned that beadboard rates very close to the extruded styrofoam in ASTM tests on water absorption, and has had many years of successful use below grade. Anyway, three inches of beadboard sloped down from the foundation and was covered with about a foot of earth for grass to grow.

The construction of the Burns house is unusual in several places. I don't like to see a lot of penetrations of the envelope air space because I'm concerned about having a very good vapor seal at the outside of the air flow space. We evolved a system of beams at eight foot centers from east to west the width of the building; the beams run north to south and they're spaced eight feet apart. That permits floor joists and roof purlins to go between beams so air flow spaces are eight feet wide instead of 16" or 24" wide. This also permits, with the eight foot beam spacing, the use of smaller joists and purlins which are only 2x6's yet are, of course, quite rigid because there's such a short span. We chose the eight foot beam space mainly because that's the maximum that you can use with 2x6's. (On other houses we have spaced beams at 12 foot centers and used 2x8's fpr those spaces.) Floor beams and roof trusses were glued and nailed on the job.

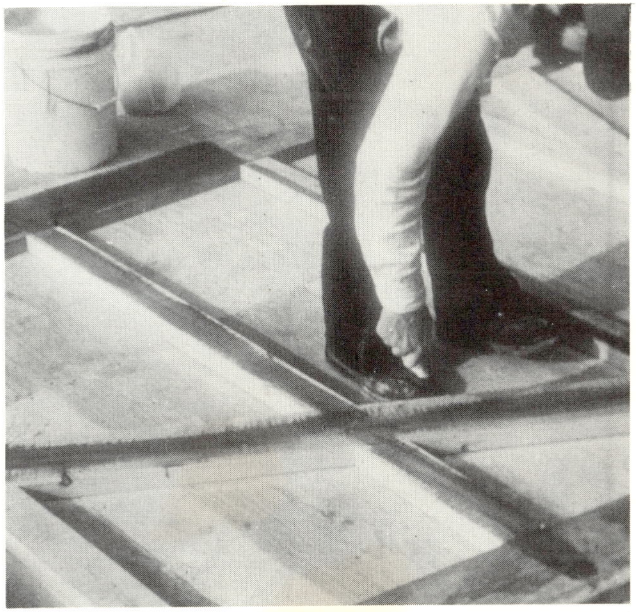

On top of the roof purlin 2x6's are 2x4's to make a 1½" air space under the roof plywood. This airspace vents moisture from the roof insulation, and keeps the roof surface cool in winter so snow won't melt.

The east wall (photos on right on 2.6) has 2x4 studs on the inside and horizontal 2x4 girts outside, with two layers of friction fit fiberglass. Outside the fiberglass here we did not use any sheathing or even any paper under the tightly-fitted vertical shiplap siding.

Polystyrene foam on the outside of the foundation is protected from physical damage by a coating of reinforced plaster (photo on 2.8). In this case we used Dryvit, but there is a less expensive system using Acryl-60, a liquid mixture made by Standard Drywall. If Acryl-60 is used with portland cement plaster in place of some of the mixing water, it makes a plaster that has excellent bonding and durability. The foam board is primed before applying the plaster by painting on a coating of Acryl-60. While it's still moist, trowel on the plaster, and press the reinforcing mesh into the wet plaster. Let that layer harden and then go back and put on a finish trowel coat.

The photo shows the north wall being plastered, at the edge of the garage. The part of the garage that is near the heated part of the house is beyond that white block of beadboard that sticks up near Bob's hand; we've got insulation across about four feet of the garage foundation. Above the stucco we used fiberglass insulation in two crossed layers.

The underside of the roof framing at the living room is shown in the top left photo on 2.10, with roof-venting 2x4's next to the roof plywood. The 2x6 roof purlins rest on a 2x4 ledger which is part of the glued and nailed roof beam. At top right the roof has been insulated and the vapor barrier installed, and gypsum board is being applied to the underside of the roof purlins. The horizontal 2x6 visible on the garage wall frame beyond the workmen will hold the ends of the ceiling 2x6 purlins, 12" below the roof purlins--creating an air flow space between the roof and ceiling structures.

At lower right on 2 10 the sheetrock is going on for the living room ceiling, and the sheetrock you see up at the top of the joists is the sheetrock on the bottom side of the air flow. We did not use a vapor barrier behind this gypsum because there are times when the temperature in the interior is warmer than the temperature in the envelope, and at those times the moisture flow and the heat flow is from inside to the envelope, and at other times when the sun is bringing the greenhouse up to 85° or so the heat flow and the moisture flow would be from the envelope toward the inside.

At lower left at 2.10 is the top of the greenhouse showing the 11" air flow space up above the ceiling with one of the 24" beams coming through and sheetrocked on both sides.

We thought it might help to have a smooth air flow in the north air flow space, so we put this little deflector (2.9 at left) over the top edge of the concrete foundation. We stopped it about a foot back from the windows so looking through you wouldn't see the edge

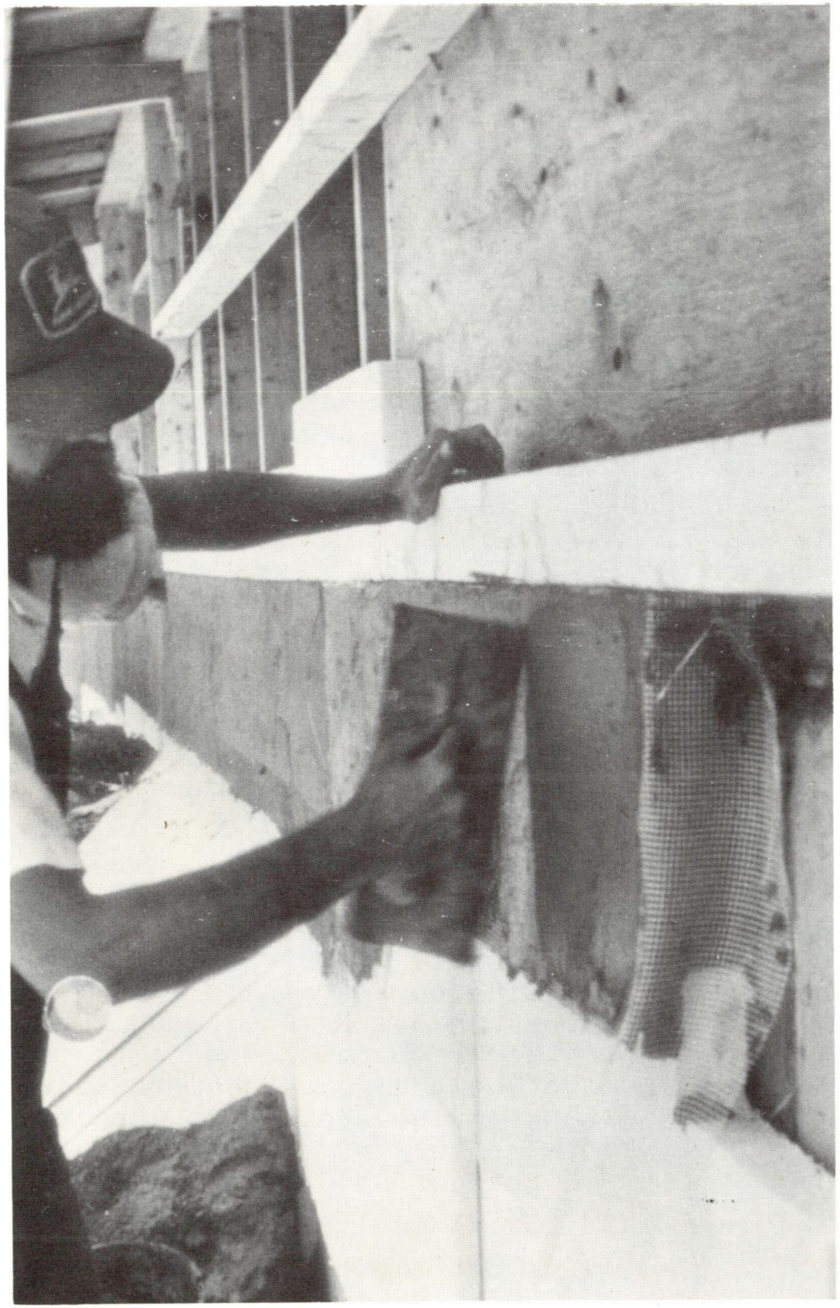

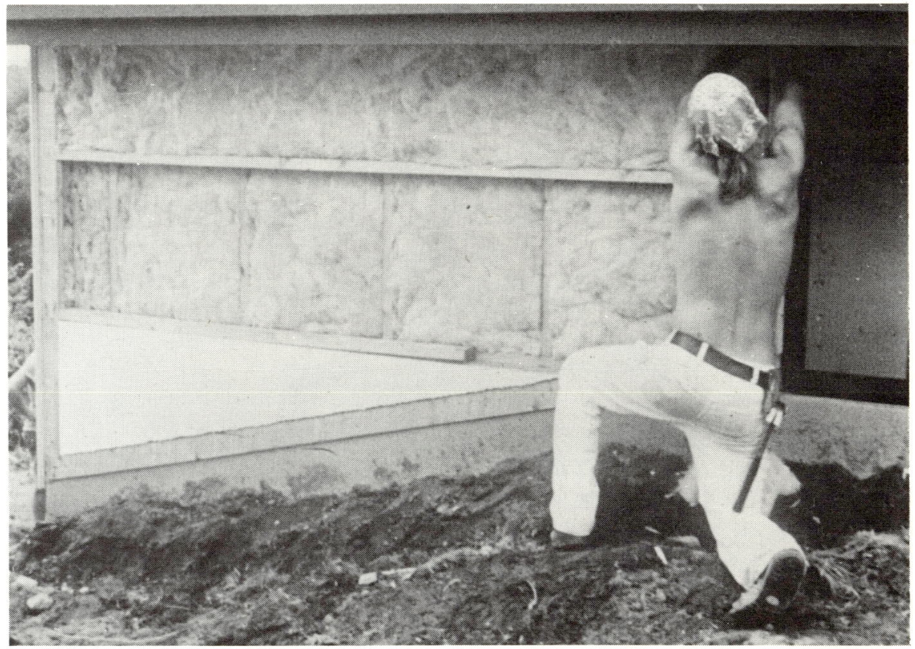

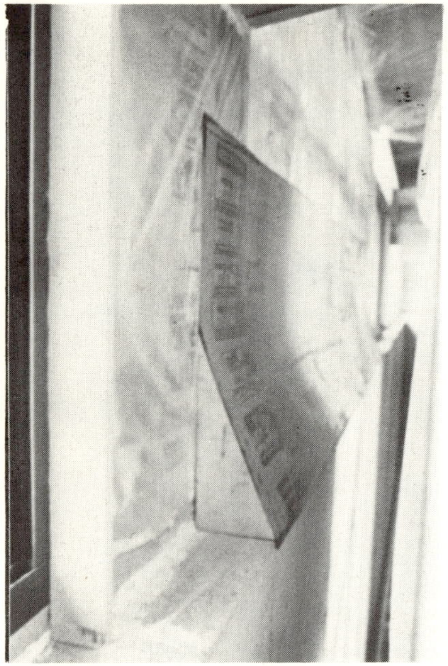

of it. We covered all the air flow surfaces with gypsum except for the sloping metal.

Looking inside through the kitchen window you see that we have a larger window outside and a smaller window inside to let the people in the house have less feeling of a thick wall. We painted the inner wall black so that as you look in through these windows you don't see anything unusual. The darker black at the top is where the sheetrock on the wall slopes back into the air flow that runs over the ceiling. We used aluminum thermal break windows because they're not affected by moisture. They cost about half as much as good wood windows and the maintenance is much less over a period of years.

The air flow from the crawl space does not flow in at the bottom of the greenhouse directly (2.11), but comes through steel grating in the narrower east and west spaces. It's much better for plants if there's a good air flow around them, but we chose not to bring the air flow through here because we wanted the deep earth connection for the growing beds.

We planned the building to use a Swedish stove fireplace that has glass doors, but the Burns wanted something they could cook on and had a stove. They haven't had a chance to do much cooking because they've burned about a tenth of a cord of wood for the year. Sometimes when the house needed heat they turned on the gas stove for a few minutes--Mrs. Burns said, "I made a batch of cookies."

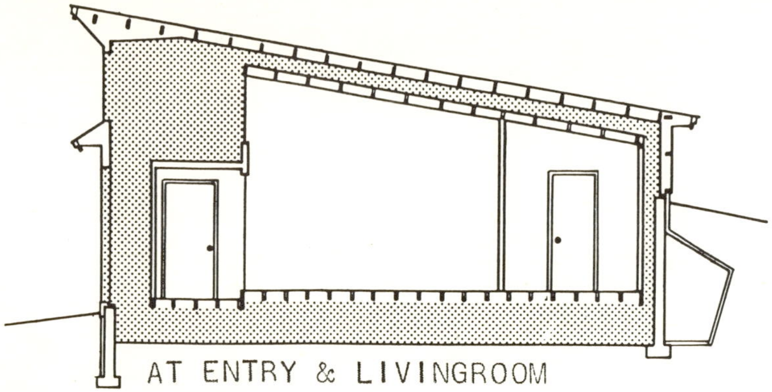

AT ENTRY & LIVINGROOM

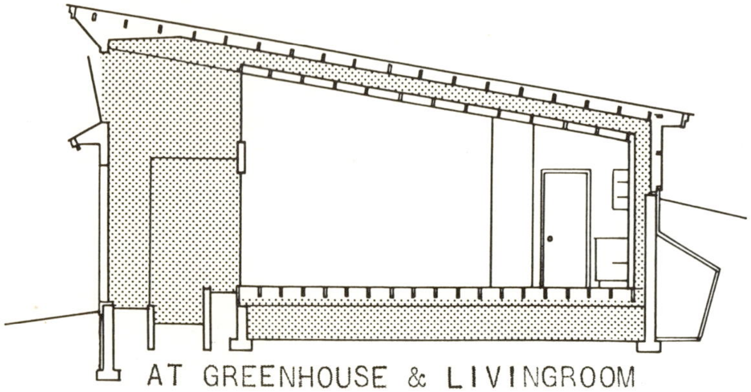

AT GREENHOUSE & LIVINGROOM

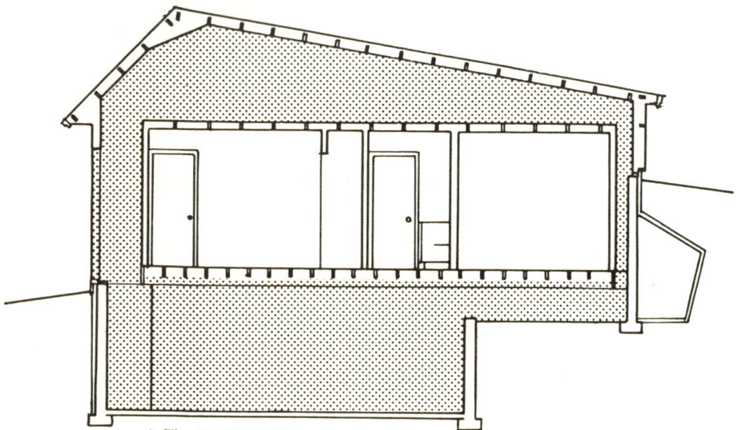

AT BEDROOMS

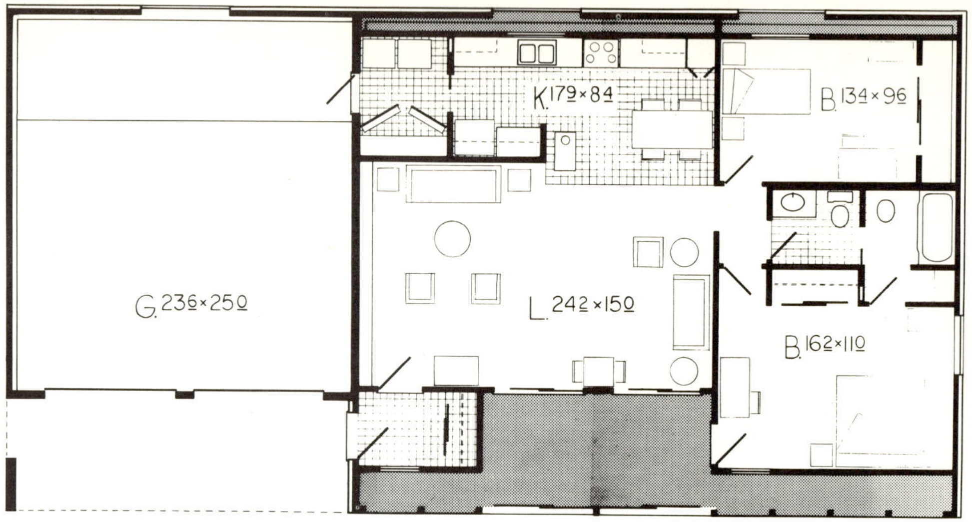

NOTE: CONTINUOUS 6 MIL VAPOR BARRIER ON ALL INTERIOR SURFACES OF EXTERIOR INSULATION; LAP & CAULK ALL EDGES & PENETRATIONS.

NOTE: USE PRESSURE TREATMENT SYSTEM WHICH IS NON-TOXIC FOR PLANTS (WOLMAN SALTS – NOT CREOSOTE OR PENTA.).

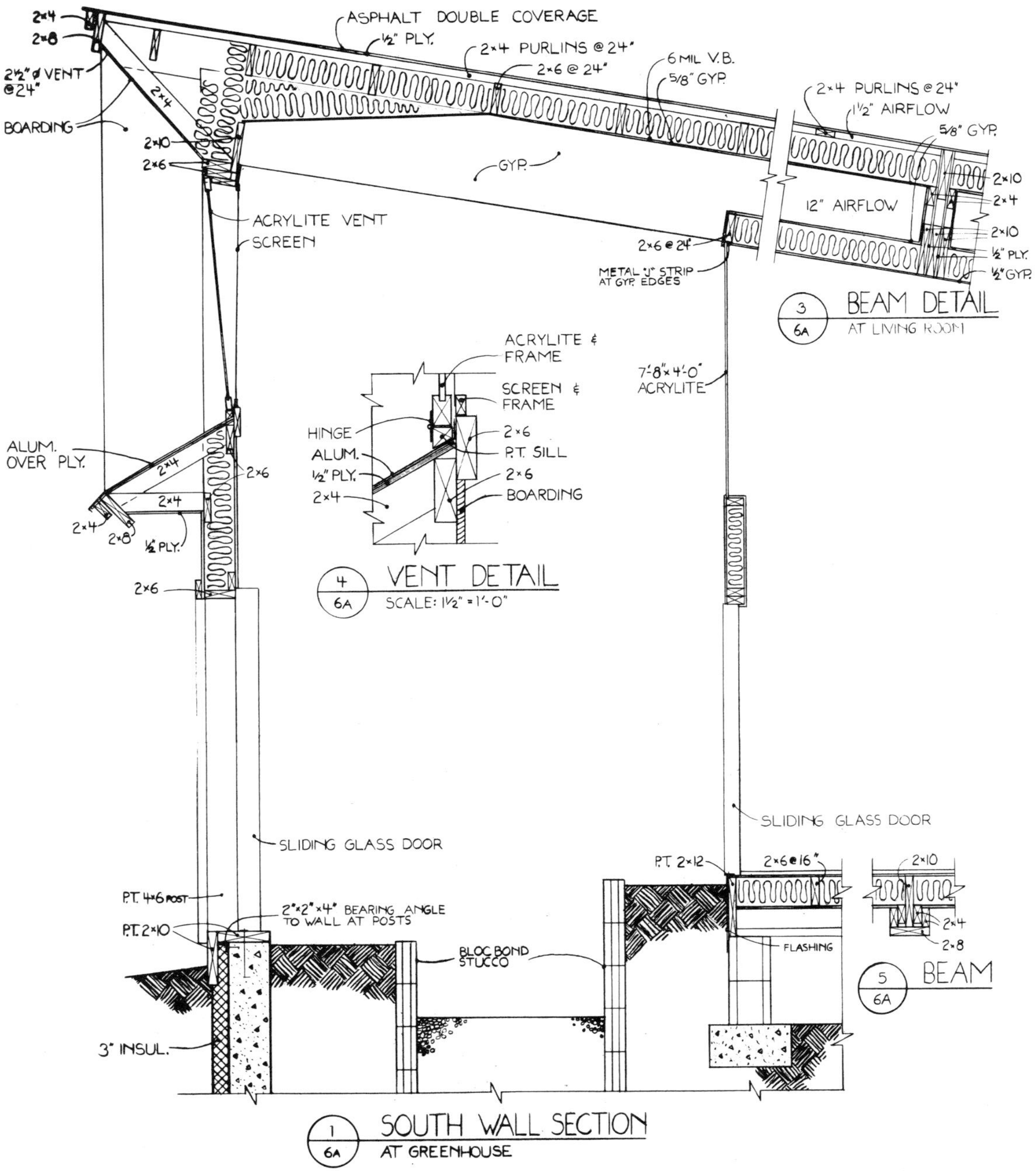

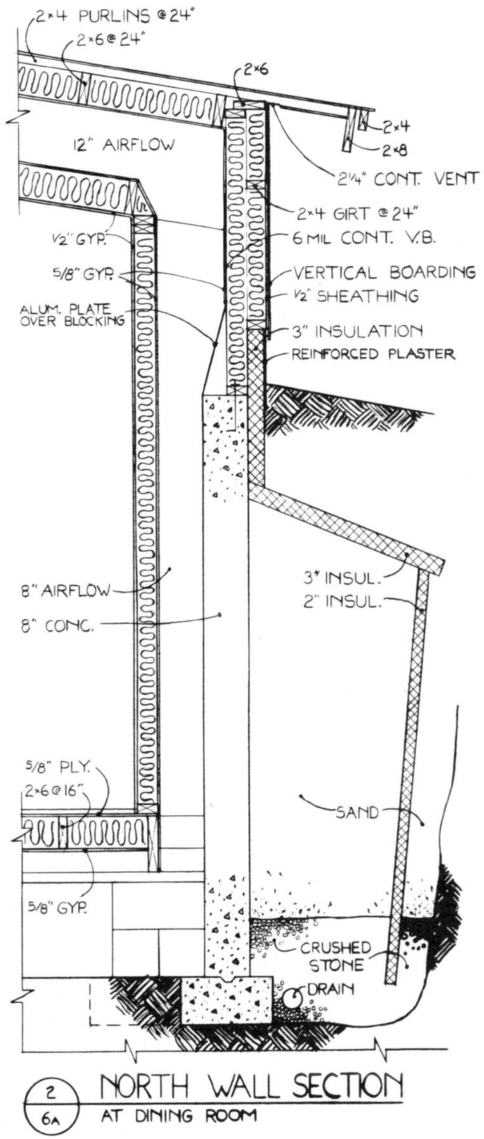

3. BEALE HOUSE

From Galen Beale's diary:

June 12 - Moved onto land and set up tent - have narrowed the house design down to an envelope - like there being no cement mass for floors or walls, a greenhouse at last, lots of light in the house without having to cover the glass at night.

June 25 - I have everything I need in three apple boxes in this tent! Don doing the house drawings - I really like how he opened up the space.

July 15 - Got the final drawings from Don - really don't understand them at all.

July 16 - Beginning to work on the lumber list which makes building seem possible. Have decided to do the contracting myself as the way I can save the most money. Tried to write a schedule for myself - two days away from the house with the kids each week, one day for making baskets.

July 21 - Fascinating conference on envelope houses at Don's Boscawen house. It's great to be able to go over and see that house in the process of being built while I'm building mine.

July 23 - Driveway in - looks great, foundation in - don't feel elation about that - just fear. This is not going to be fun.

July 29 - Finally decided on a builder - foundation to be poured tomorrow, then blocks - mason sure is skeptical about the foundation. Trying to hire as much done locally as possible - feel better now that I can see progress, but the house sure is little. Have tried to make a list of what I will worry about and what I will not. I will not worry about November, fancy food, a social life, I will worry about the kids and the cost of the house. So many decisions to make - all biggies - trying to narrow it down and eliminate some options - no plywood, plastic, metal.

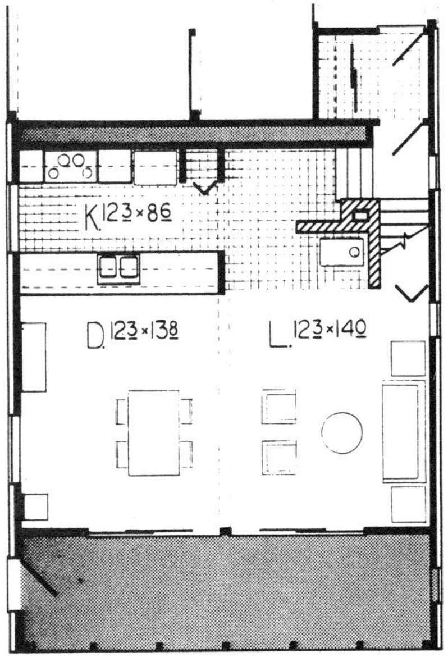

FIRST FLOOR

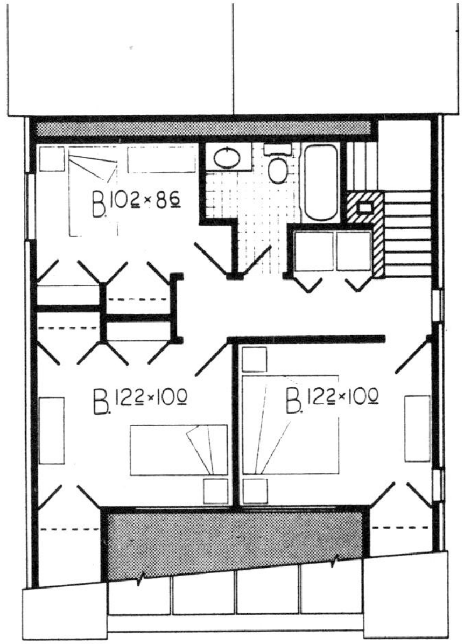

SECOND FLOOR

<u>August 6</u> - Have been hauling lumber from Minery's sawmill in the unending 90^0 heat, trying to get ahead before we start - trying to pile it logically. House construction starts tomorrow.

<u>August 11</u> - Working like a dog! The lumber list was way off, given it up, just guessing now, takes a lot of time, everyone trying to figure what we need tomorrow each day but the crew is great and very relaxed. Guess this is the way it's going to be. Spent most of the very long week chasing 2x8's - struggling to get the materials here - millions of trips.

<u>August 14</u> - Crew off for two days again because I can't keep enough lumber here - we're working 6 am to 6 pm which is a great way. But I really wish I could just stay home all day and work on the house and someone else would hassle with the stores. A phone today - that will make life easier. I put up a mailbox - getting territorial.

<u>August 24</u> - House is coming along. I know I'm doing it the hard way. Rough lumber very uneven - takes lots of time. Shopping for everything, endless hassle about materials - don't think the crew has worked a full week yet. They are getting impatient at my disorganization - can't blame them, but I do seem determined to do it my way. Going to start the barn tomorrow to keep the crew busy while waiting for the roof trusses.

<u>September 5</u> - House stalled again. Sawmill broke down. School started, the summer seemed so short. Fear the crew will quit - they have a point. Second floor walls up. It's a nice house.

<u>September 15</u> - Crew quit. Trusses cancelled because of high winds. House and barn all framed for $12,000. I wonder who I'll get - creeping panic!

<u>September 18</u> - Found another crew.

<u>September 20</u> - Trusses finally came - they're wrong! Not enough and the pitch is wrong. New trusses will take another week. All stopped on the house for two weeks now, trying hard now to keep remembering the things I am not going to worry about. First frost.

<u>September 21</u> - Went sobbing over to Don's, told him I was beginning to take the trusses personally, he said I should read a book called <u>You Are Not The Target!</u> We ate some wild grapes and discussed the house - so calming and nice. Thought it would be ok to board in before the trusses come. Inexperience makes you overcautious and slows everything down but it's cheap.

<u>September 27</u> - Getting cold - have to figure out at this point how to best spend the rest of my money - hire yet another crew to do the roof? Going to plywood the roof to save time. Moved the tents into the barn, good to get out of the wet in the morning.

<u>September 29</u> - Trusses came. House looks huge now. Crew coming to do roof. Now quite used to carpenters coming and going, it's nice while they're here.

<u>October 5</u> - Going to do it - Move in! Stuffed the envelope with fiberglas and got a space heater. We're all delighted to be in - the kids say it's great now because there are no

more worms in the tent - what little troopers they have been. It's much colder in the house than in the tent and feels so closed in - can't hear any of the night sounds.

October 17 - Snow. Don Booth's crew coming - very professional - Don comes each morning and I'm amazed at how much of what he does that I did. Roof on, greenhouse glass finally in. Did it with the help of strong friends - the glass company brought the glass (heavy) and suction cups to hold on with - I couldn't even watch. Got away for a night but I have this anxiety if I'm gone for long that eats at me, I always want to be back here working on the house. I have not kept to my schedule of getting away with the kids, it's been all-consuming.

October 30 - Sheetrock tomorrow. Got bricks from woods and piled them in the house to keep warm - what a mess - they are old and the mason hates them because they're so uneven.

November 13 - Sheetrockers really here AND carpenters AND a new mason!

December 14 - Taking time now to enjoy the house. Sitting, reading, watching the sunsets. Having so much glass on the south is wonderful, makes you just want to sit and look at the view without feeling you should go outside. The house is always so light - I love it. Planted the greenhouse with Boc Choy, Chinese cabbage, lettuce, spinach - it never needs watering it seems. I shovelled all the dirt for the earth beds in by hand, had a loader dump the earth at the door and now I don't see a need for the wide aisle for a wheelbarrel - takes up a lot of space.

December 22 - Going away for Christmas - it's a whole new way of thinking to just walk away from the house and not have to find someone to come every day to stoke the woodstove. Still lots of leaks in the house, needs cauking, but it's so pleasant to live in with the high humidity and so much glass that a day can never look dreary because you are always looking through the greenhouse and this incredible green color of whatever is growing there. I wish there were a window on the north so I would have some warning when people come in the driveway because it is very hard to hear with the house this tight - and a sliding door instead of a window on the second floor because it is so much warmer in the upper greenhouse it's a great place for hanging plants, I am continually climbing out the window. I love heating my house with the sun and the house just doing it by itself - no machines to fiddle with, no rushing around at night to button up the house. To wake up in the winter with the ice on the windows and to feel the sun come up and eat away at the ice and turn it into pools of water in the greenhouse and then gradually raise the temperature of the house, it's a wonderfully natural process.

Galen Beale - house and barn expenses:

Labor	$10,419.99
Materials, plywood, nails, insulation, hardware, roofing	4,324.65
Wood	3,395.69
Driveway, grading, backfill, well, septic	2,680.82
Rigid Insulation (Dow)	868.00
Greenhouse Glass	1,300.00
Sheetrock	1,210.00
Trusses	825.00
Chimney, materials, labor, stove	974.00
Bio-Loo	845.00
Foundation - house, barn, labor & materials	901.15
Windows, doors, stainglass	2,663.75
Electrical	200.00
Plumbing, labor, materials	876.24
Public Service, burying line	107.25
Total	$31,492.28

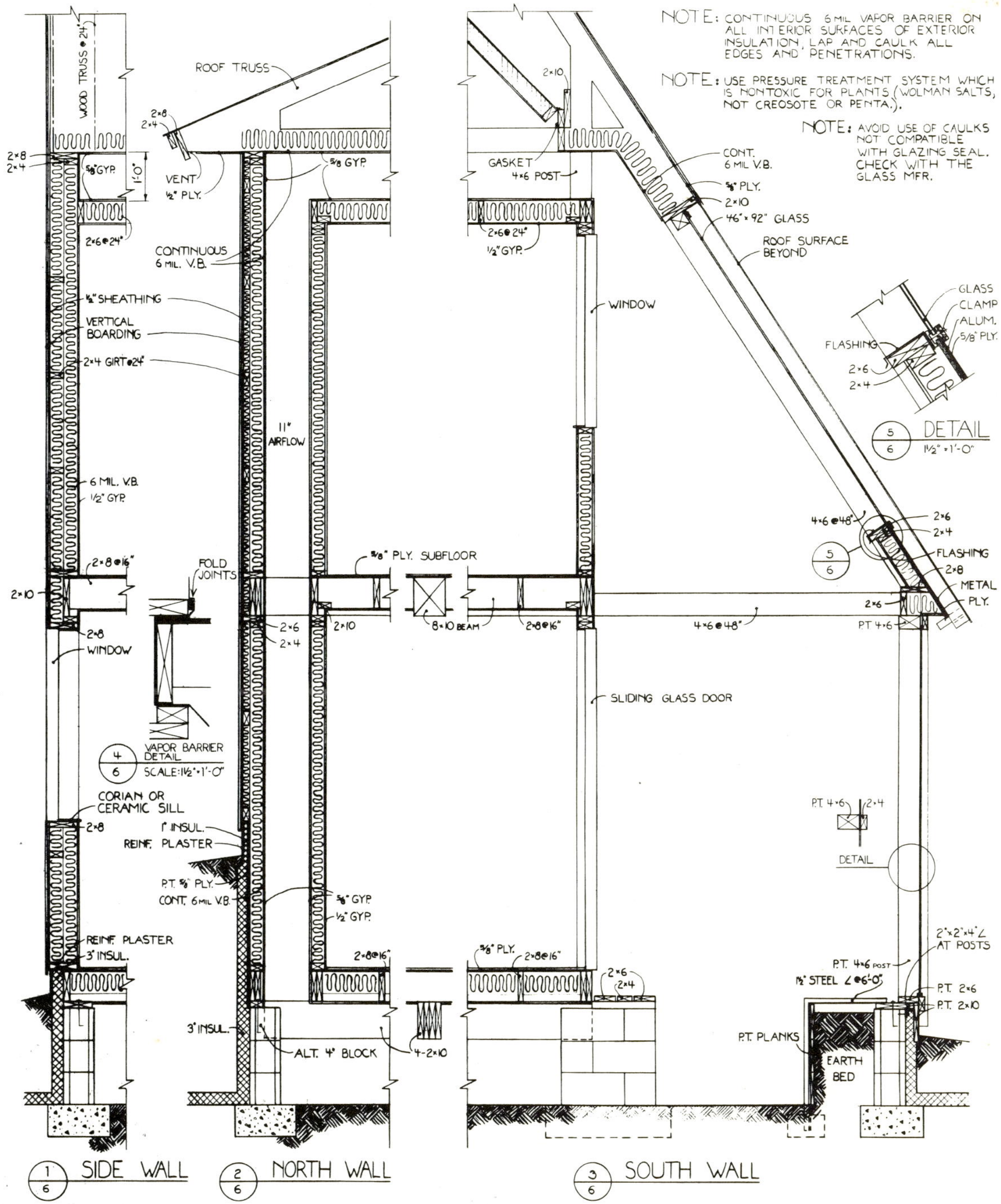

4. THREE HOUSES by HANK HUBER

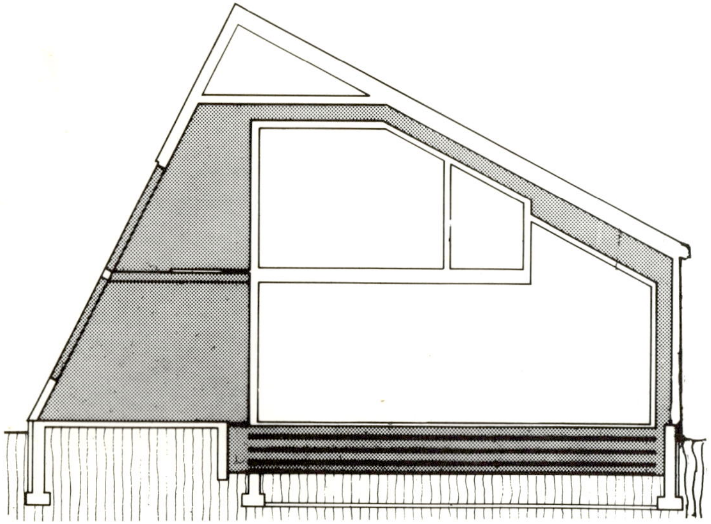

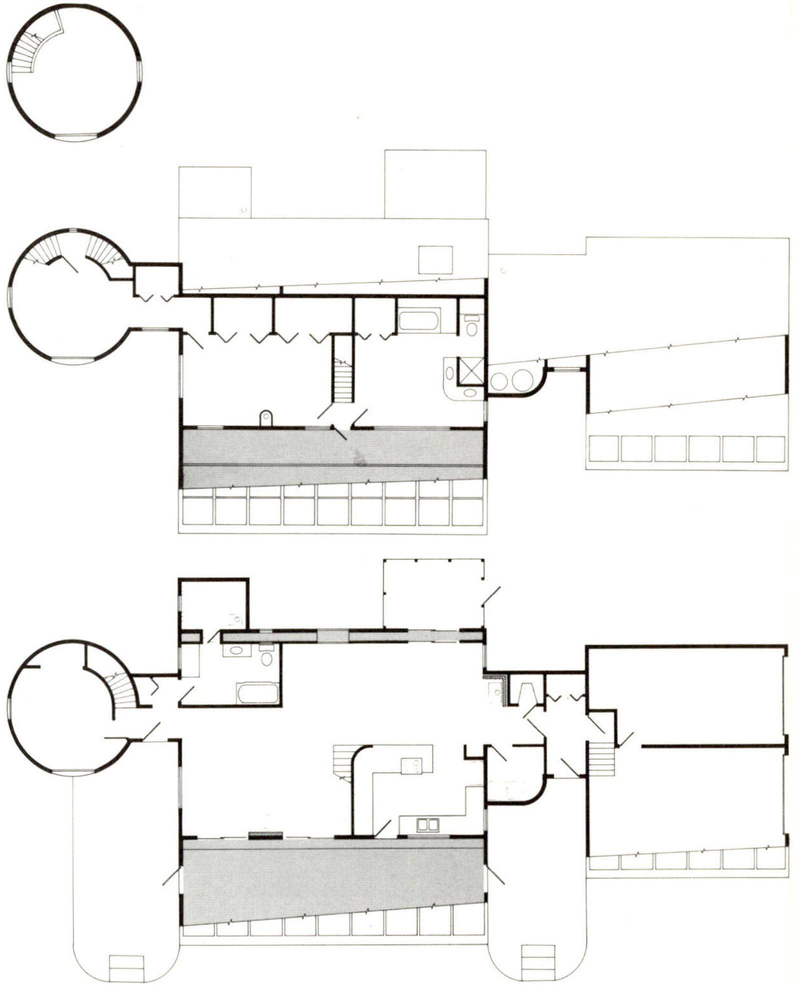

HUBER-WOOD HOUSE Comments by Hank

My long standing concern for the growing energy and food crisis led me to design this house during 1976-1977. Construction and occupancy occurred in 1978. The envelope system was a good approach at integrating a large food producing solar greenhouse with a thermally efficient residence.

The envelope portion of the house is a 36'x 36' modified salt box with a two story greenhouse on the south side. A three story silo structure, not in the envelope, is solar tempered with 25 sq. ft. of glass per floor and also heated by gravity convection from the main part of the inner house. Warm air to the silo flows through a series of large floor registers and stairwells. A passive solar heated workshop/garage occupies the eastern part of the structure. Three of six glazing panels here will have domestic hot water collectors installed inside the glass.

Initial thermal calculations I had done indicated a need for greater area for heat transfer in the crawl space. My solution was to place 2,100 12" two core concrete blocks over compacted earth just below the main living floor. The cores of the block were lined up to provide minimum restriction to air flow. Monitored performance has shown a considerably higher sub-floor heat storage here than in conventionally built envelope homes. The perimeter of the four foot foundation wall was insulated with 3" polystyrene foam plus a protective layer of 4 mil polyethylene.

The wall and roof system was conventionally framed with inner ceiling and north wall added after the structure was weathertight. The outer roof is 2x12 on 24" centers. The outer north wall is 2x8 on 24" centers. The inner ceiling and north wall are 2x4 on 16" centers. The roof plenum varies from 9" to 24" wide allowing access. See upper right photo on page 4.4. The north wall plenum is 11" wide.

The silo was framed in the round with 2x6 on 16" centers and diagonal 1x3 strapping (see photo on page 4.4) on 16" centers. Exterior vertical shiplap siding insured a very rigid durable structure. Plywood was used only on roof and floor decks.

North wall insulation is R26 outer and R13 inner. Roof is R30-40 outer and R13 inner. Remaining house walls are R19 with an additional layer of 1" polystyrene foam on the inside of the 16' diameter silo walls.

The 12'x36' greenhouse is glazed at a 24:12 pitch with 18 panels of 46"x92"x1: tempered insulating glass set into a rabbeted 4x6 Douglas Fir frame. The frame was treated with three coats of clear Cuprinot prior to glazing. A long lasting weathertight glazing system is shown on page 4.10. Note the aluminum capping for the wood framing. The choice of sloped glazing was one of aesthetics and concern for adequate overhead illumination for year round plant growth. Summer venting is accomplished through a screen door and two trapezoidal screened vents on east and west ends of the greenhouse. Additional venting is provided into the gable space. Greenhouse temperatures are about 5°F warmer on a sunny summer day than those directly outside.

Venting works well enough to eliminate the need for summer shading. Opening north windows and sliders to the greenhouse creates a natural convection through the house. While summer sun penetrates most of the greenhouse, it is excluded by internal overhangs from penetrating the inner house, thus maintaining a remarkably cool indoor temperature.

The silo represents a thermal drain on the envelope accounting for about half of the 1½ cords of supplemental fuel required each of

the last two winters. The few installed baseboard heating units have never been used.

The higher relative humidities experienced in envelope houses (or any other designs employing large actively used greenhouses) suggested the omission of plywood as a wall sheathing. With a perm rating of 1, exterior plywood is classified as a vapor barrier. Higher relative humidities require careful attention to quality and placement of vapor barriers. The proper location is close to the inside of the exterior shell. A minimum of 6 mil polyethylene should be used. Vapor barriers should not be used on the inner shell, thus allowing a free exchange of moisture in both directions between the inner house and the envelope.

Addressing the problem of the energy crisis requires our attention not only to space and hot water heating but to the local production of food. The passive solar greenhouse can make a substantial impact on regional food production. In combination with summer gardens, solar coldframes and root cellars, many families have the potential of becoming food self-reliant.

Our greenhouse has provided Swiss chard for the last 13 months from a single plant, tomatoes in January, parsley, cabbage and broccoli in February. See page 4.4. The minimum greenhouse temperature in 1979-1980 was 40°F without supplemental heating or night curtains.

PETERBORO, NH HOUSE

Lower left at 4.6: Active growing in a passive greenhouse. 3½x15" growing bed just beginning operation in January 1980, one month after owners occupied building.

Upper left at 4.6: An interior photo showing the open concept kitchen, dining, living area. Four foot square flow-thru skylight (see detail) provides overhead illumination for the kitchen.

Thru-flow Skylight Detail

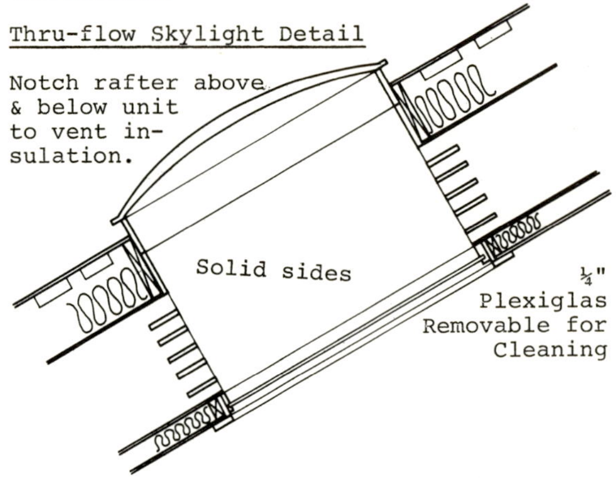

Upper right at 4.6: This is a southeast view as the house was nearing completion. Entry airlock is east of house envelope portion. Double gable vents provide about 70 sq. ft. total upper vent area. Lower venting occurs through east/west sliders and insulated vent doors below each of vertical glazing units (see louvre/vent detail). Mounting panel for domestic hot water collector is seen behind top row of south glass. A sheltered east deck provides a cozy outdoor place for cool weather sunbathing.

Lower right at 4.6: Greenhouse interior at Sharon house showing slatted grating for air flow. A very open grate, made from southern pine, is located on the south side of the greenhouse floor. Brick pavers comprising rest of floor are seen at bottom of photo. Vent swings in. Door is well-insulated. Insulated vent doors in Peterboro house use Andersen flexivent basement window hardware, allowing complete removal of unit for summer venting.

PETERBORO HOUSE

I designed and built this three bedroom envelope in 1979. Living space is 2,100 sq. ft. including the first floor of the greenhouse. It has a full basement with 3" outboard polystyrene insulation to increase the available thermal mass. Site analysis indicated substantial flowing water below grade which would tend to drain heat away from the basement store. We, therefore, included a 2" layer of polystyrene foam under the basement floor slab. A simple high static water table would not require this extra insulation. It is much better thermally to have direct connection to the earth as a low grade heat source for the envelope.

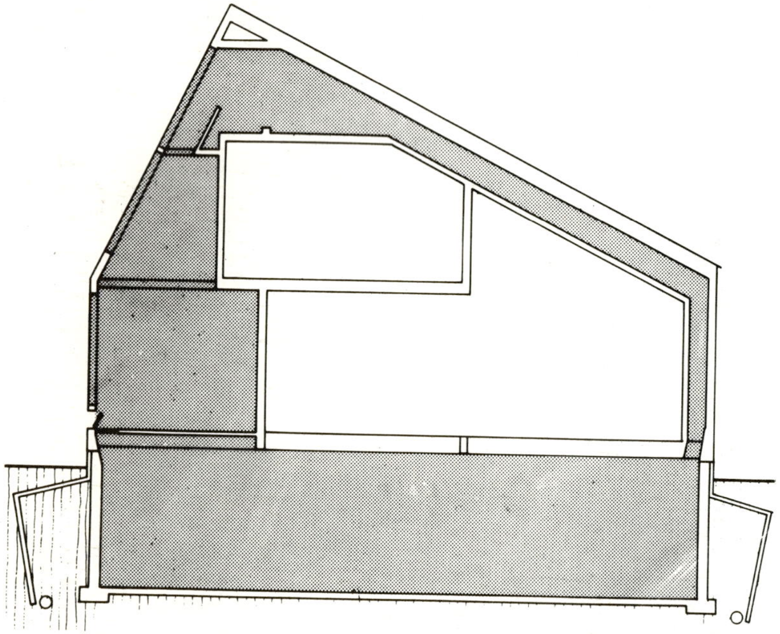

Insulation values are R19 exterior shell and R13 interior shell. The net south glass area is 690 sq. ft. including about 100 sq. ft. for an integral domestic hot water collector. See detail at 4.9. Envelope width is 18" ceiling and 12" in the north wall. Gable and vents with four insulated, remote operated doors provide for summer venting.

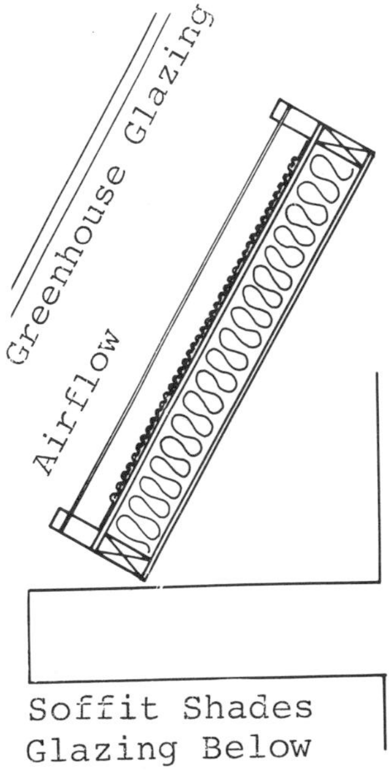

This insert corrects omissions on pages 4.7 & 4.8 of The Double Shell Solar House.

PETERBORO HOUSE

I designed and built this three bedroom envelope in 1979. Living space is 2,100 sq. ft. including the first floor of the greenhouse. It has a full basement with 3" outboard polystyrene insulation to increase the available thermal mass. Site analysis indicated substantial flowing water below grade which would tend to drain heat away from the basement store. We, therefore, included a 2" layer of polystyrene foam under the basement floor slab. A simple high static water table would not require this extra insulation. It is much better thermally to have direct connection to the earth as a low grade heat source for the envelope.

Insulation values are R19 exterior shell and R13 interior shell. The net south glass area is 690 sq. ft. including about 100 sq. ft. for an integral domestic hot water collector. See detail at 4.9. Envelope width is 18" ceiling and 12" in the north wall. Gable and vents with four insulated, remote operated doors provide for summer venting.

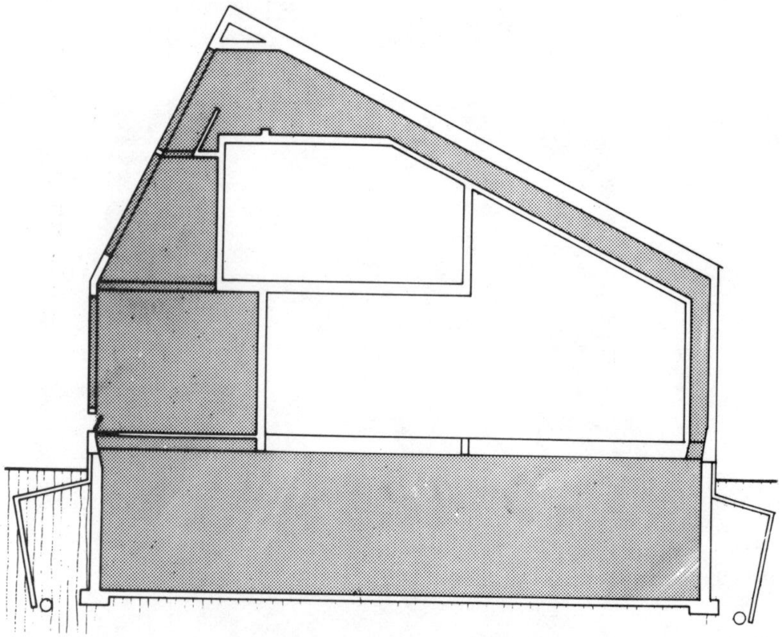

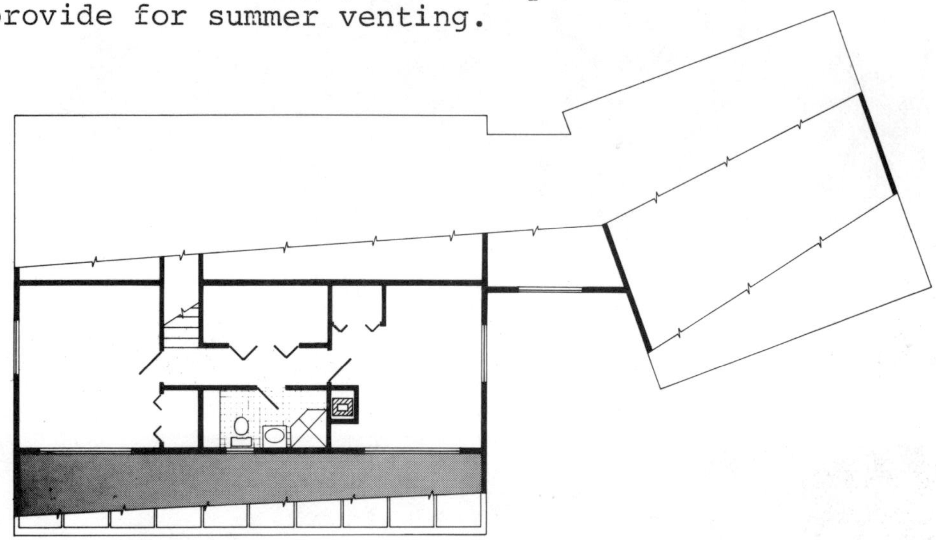

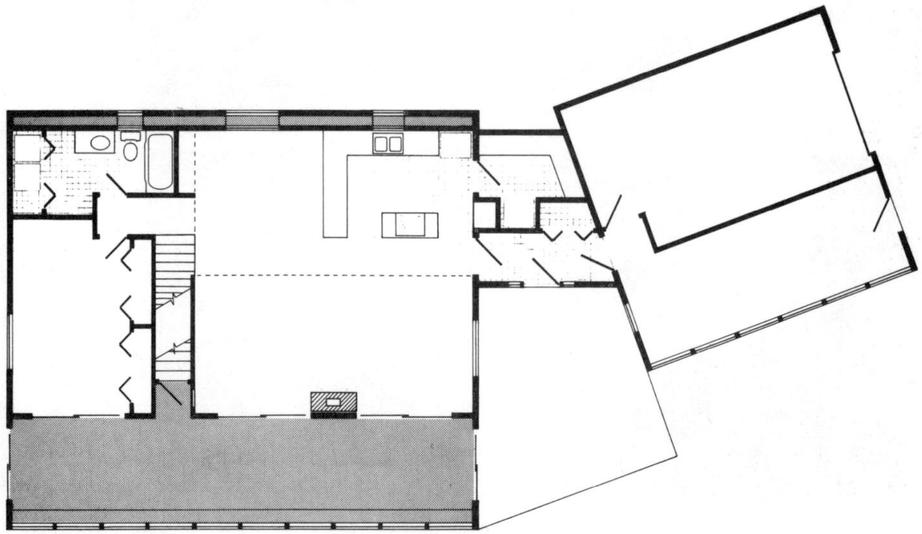

This insert corrects omissions on pages 4.7 & 4.8 of <u>The Double Shell Solar House</u>.

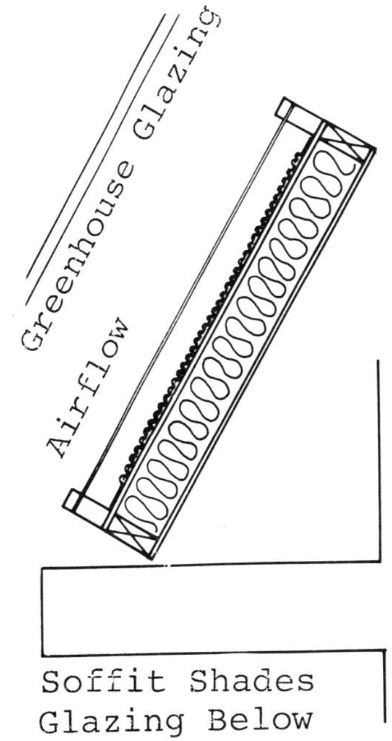
Greenhouse Glazing
Airflow
Soffit Shades
Glazing Below

SHARON, NH HOUSE

This house was also designed in 1979. Ned Eldredge of Peterboro was the builder. While it employs a similar style and construction technique, its floor plan provides about 2,600 sq. ft. of living space. It has a net south glass area of 621 sq. ft. and also includes a 100 sq. ft. domestic hot water collector above the second floor level. Both houses employ a sliding pocket inner north window and double glazing casement exterior windows (see detail). A fan sided convection system (see detail) was installed in the Peterboro house during February-April 1980 to compare the thermal performance of these two houses. Some results are presented in Chapter 7. Fuel consumption for both houses was less than two cords of wood for the winter of 1979-80.

Lower right at 4.8: Kitchen window showing airflow path between exterior casement unit and sliding pocket window on the interior.

Upper right at 4.8: Upper level of greenhouse showing 100 sq. ft. domestic hot water collector. One layer of Kalwall Sunlite Premium covers Tube Mat collector plate. See section detail. System is active with pump and tank in basement. Differential thermostat actuates circulating pump. System produces about 75% of hot water needs.

Upper left at 4.8: Living space on two levels of 2650 sq ft, 650 sq ft glazing with 100 sq ft domestic hot water collector located inside greenhouse. Large gable vents on east and west. Excess heat from greenhouse is dumped into gable by opening insulated horizontal trap door. Six foot sliders on east and west of house allow access and venting. Vents under each vertical panel provide additional venting.

Lower left at 4.8: Interior living space, country kitchen looking west toward living room. Two sliding doors allow access to greenhouse.

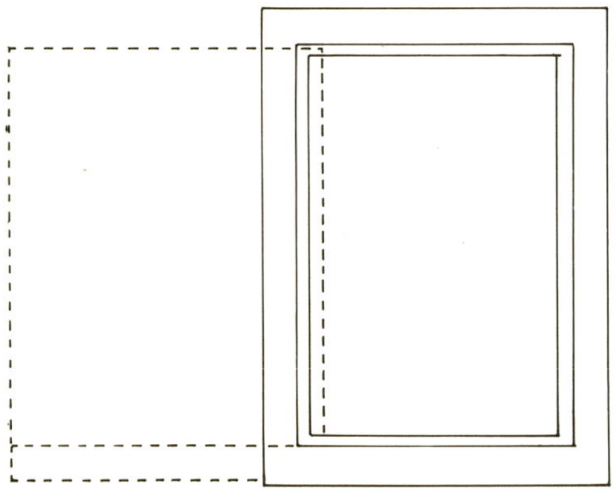

2x4 Nailed to Inside Studwall Supports Retracted Glazing

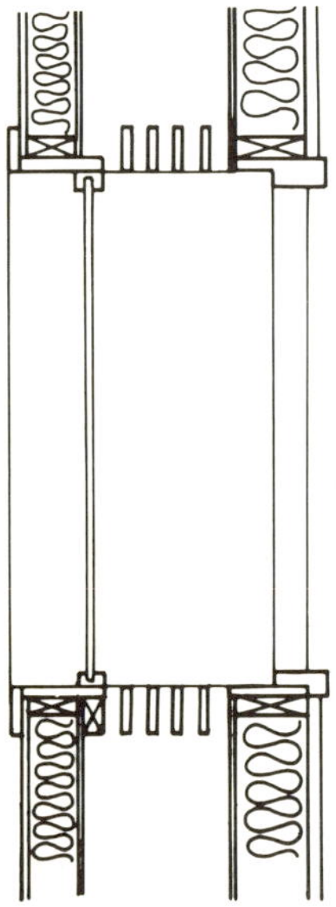

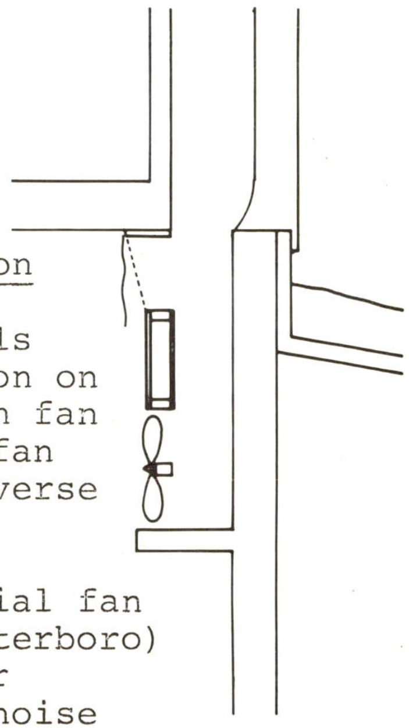

Fan-aided Convection

.002 Poly flap pulls into dotted position on hardware cloth when fan is on, drops when fan is off to allow reverse convection

20" circulating axial fan (3 installed at Peterboro) mounted to wall for minimum vibration noise

SHARON HOUSE

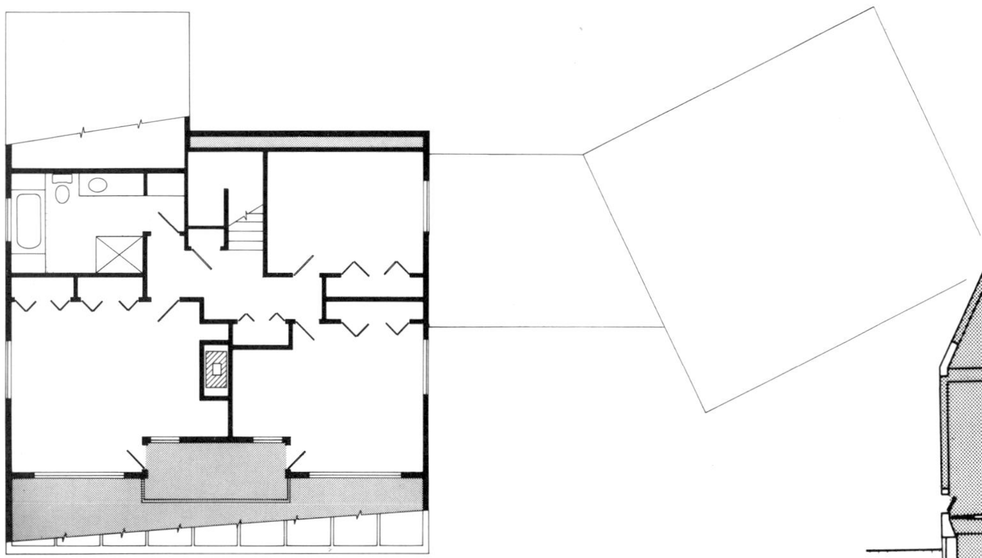

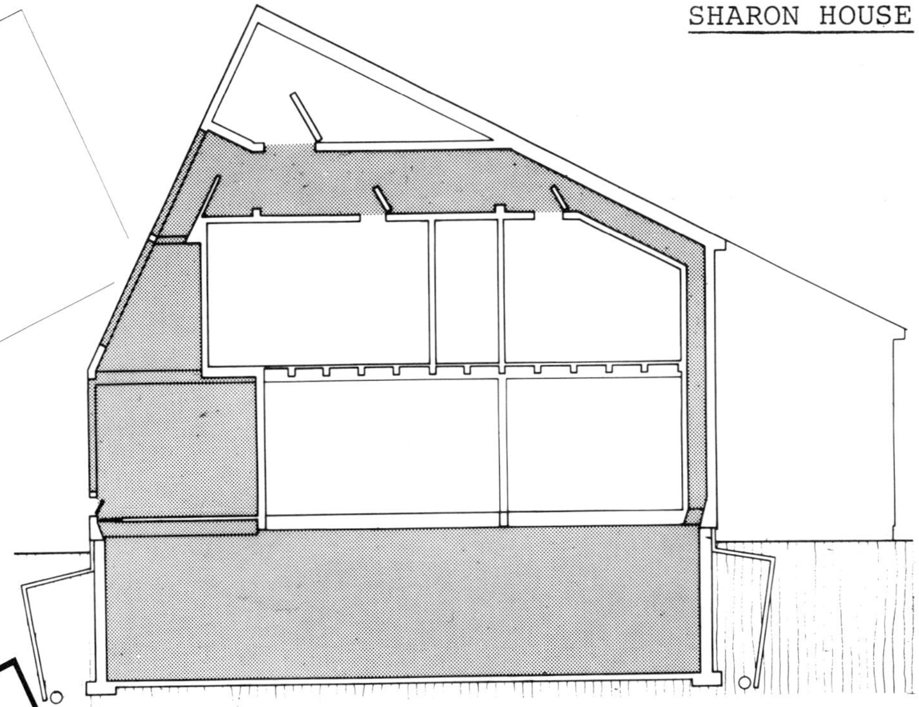

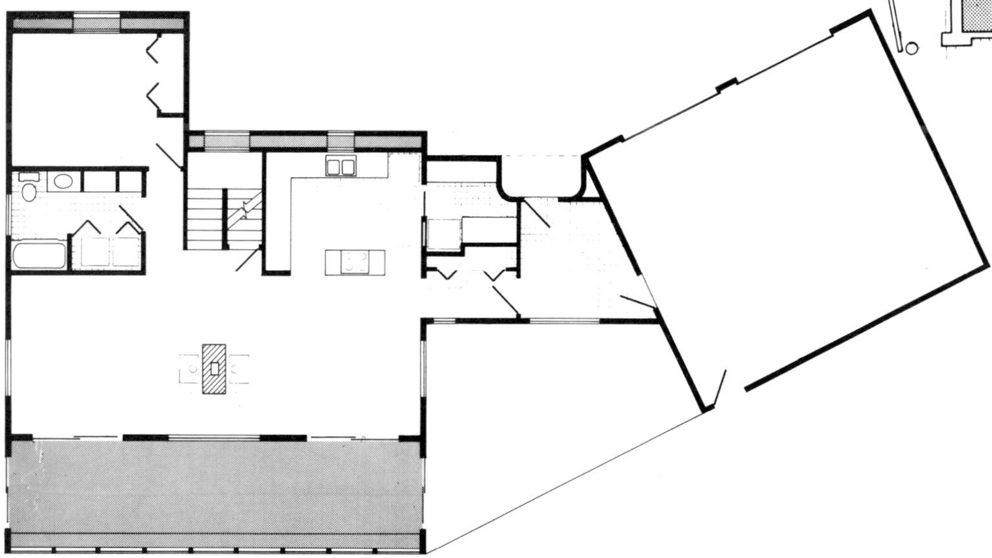

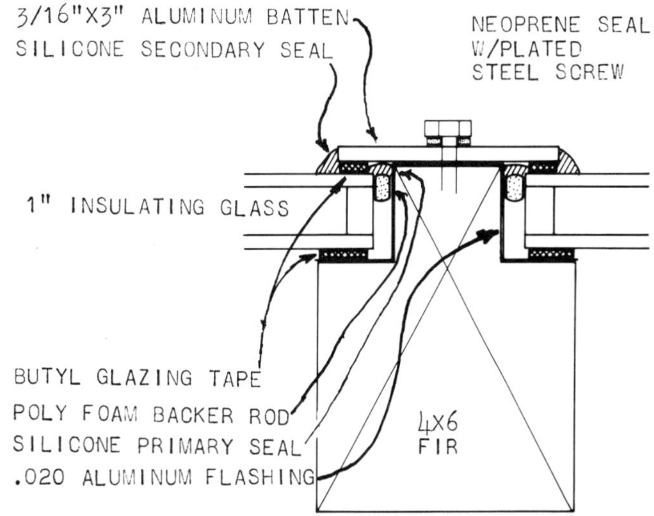

5. Discussion on Design

by Bob Henninge

In 1979 and 1980 three envelope house conference/seminars were organized by Community Builders to bring together designers, builders, and owners of envelope homes to share their experience and understandings of the envelope house: how it works, why it works, how to design and build it, what it is like to live in it. This section of the book pulls together and presents in an orderly form the insights and information shared during the conferences. The material is presented in question-and-answer form, organized as follows:

I (Pgs. 6.6-6.9)
Insulation and Vapor Barriers; types of insulation and their limitations, insulating foundations, the importance of a continuous vapor barrier, and techniques for achieving one.

J (Pgs. 6.9-6.11)
Fire Protection; approaches to fire safety, building and installing fire dampers, reactions of local fire officials.

K (Pgs. 6.11-6.13)
Air Quality issues; odors, internal pollution, humidity, and the use of heat exchangers.

L (Pgs. 6.13-6.14)
Other Special Concerns about protecting the internal environment from condensation and pest damage; noise levels inside.

M (Pgs. 6.14-6.16)
A survey of the current Costs of envelope houses; discussion of strategies to hold costs down.

A. GENERAL THEORY

A-1 What is the envelope concept and where are its roots?

The technique of the buffering envelope is described succinctly in the beginning of Jim Berk's "Working Papers on Convective Loop Housing": "Enclose the inner living region within an outer living region where the contained air in the outer region is convection circulated up, over, down and then under the living region. Heat intake on some surfaces and heat losses on others all work together to drive the air around the inner living region, and the heat into and out of storage, both in the structure and the earth beneath the house." While it sounds like a remarkably simple idea, its heritage is diverse and includes many awarenesses from both the passive solar and energy-efficient design traditions.

The envelope house employs most conspicuously the heat-producing greenhouse, and the natural convective loop for heat distribution. That greenhouses can be built so as to guarantee, through the provision of a large thermal mass for heat storage, an unfreezing environment for growing plants year-round, is something that has been proven many times in many climates. One simple model of an envelope house is that of a normal house built inside a solar greenhouse, with the convective flow distributing heat all around it so that the inner house itself provides much of the mass that moderates temperature extremes in the greenhouse (loop area) through the day-night cycle.

The most general equation for conductive heat loss through a membrane is Heat Loss = $\Delta T/R$, where ΔT is the difference between the temperatures on the inside and outside of the membrane, and R is the insulative value of the membrane. Designers of energy-conserving houses have taken one or both of the two possible steps to minimize heat loss. By concentrating on increasing R, the resistance to heat flow, they have arrived at the super-insulated house; by snuggling the house into the even-tempered earth they have reduced ΔT directly, thus effectively moderating the climate against which an earth-sheltered house seeks to maintain comfort inside. The envelope house borrows substantially from both these approaches. All walls which directly separate the internal living space from the outside (commonly the east and west walls) should be "super-insulated", and the envelope space itself reduces the heat loss of the inner house walls by bathing them in a blanket of air that is being continually warmed to the temperature of the earth mass under the house.

A-2 What's the difference between an envelope house and a super-insulated house with an attached greenhouse?

BH- In a superinsulated or conventional house, an attached greenhouse supplies heat to and draws heat directly from the living space. In an envelope house the greenhouse supplies heat that tempers the air surrounding the living space, thereby reducing heat loss. So, a super-insulated house with a greenhouse would be a solar-heated house; an envelope house is more a solar-insulated house.

A-3 How cold will the inner house get with no auxiliary heat?

DD- During construction with no one in the house, we went for about five days of overcast skies, outside temperatures of about 10°, absolutely no heat on in the house, and the interior of the space was 45°.

DC- When the Burns went away on vacation for a couple of weeks, the house went down to about 48° at night.

HH- I went away for seven days in January, and it was cloudy for five of those seven days, though not particularly cold. Our house got down to 48°, with the envelope at 42°. That compares to an envelope temperature in the upper 40's over a period of time when we were residing in the house and interior temperatures in the 60's with a little supplemental heat or in the high 50's with no supplemental heating and the same kind of weather conditions.

DD- The reason, as I see it, is that you have this source of 45° heat in the earth supplying heat to the house. If you didn't have it and you had absolutely no sun, the house would fluctuate around the average outdoor temperature. You have a constant supply of heat coming from the earth into a well-insulated shell, so the house is going to maintain the average temperature of the earth.

A-4 Is the envelope a solar-tempered or an earth-tempered house?

- My first thought when designing an envelope house was that its main function was to gather solar energy and store it into the earth mass. After my first house I realized that wasn't really happening. Today I'm hearing figures of 20% and below of the energy hitting the glass making it down into the storage area, and that seems like an insignificant amount considering the thousands of dollars it takes to put 5 or 6 hundred square feet of glass in a building. In my belief the houses are really "geothermal" houses and not houses run by the solar energy hitting the glass on the building. They are "geothermal" houses because the mass underneath the building is tremendous and contributes huge amounts of energy with small changes in temperature. What you are really trying to achieve is to envelop the inner walls of the house with an environment which is very similar to the temperature in your refrigerator--that is, 40 to 45 degrees at the lowest. Your loss through the walls is the direct result of your delta-T which is 20 to 25 degrees on the high side, probably even down to 5 and 10 on an average. Those kinds of losses you can make up very easily with internal gains; the energy comes from the compressor of your refrigerator, your stove that's running, lights you have on normally, and your own body heat. I believe you can put these houses in a forest and still save a significant amount of energy.

BB- Two years ago at the Second National Passive Solar Conference Lee Butler spoke before anybody else knew about envelopes. I was reading in the papers about six months later, after I found out about Tom Smith's house, and here was Butler saying that of all the variables in an envelope house, the one that can be taken out without killing the performance is the south-facing glass. Now when I read that at that time I said, "That's a misprint--somebody just didn't write that down right." Then about six months later some other article by Butler said the same thing; I said, "Well, that's not a misprint--this guy's crazy. How can you build a solar house without south-facing glass?" The more we looked at these houses, the more we realized that the earth was contributing tremendously to the performance of the house because of this lower delta-T value, and it is a more crucial factor in the performance of the house than the south-facing glass.

DC- On a daily basis the Burns house is functioning as a solar collector with a calculated efficiency of at least 40%. During the 8 hours of good sun, you are losing about half of your solar gain back out the glass and roof; the rest of it is going into the structure of the house, which is helping carry it through the night, and into the earth below. So I see that a different design strategy could make it just an earth-tempered house, but it certainly can also be a solar house in the sense that the sun is driving the loop in the daytime and storing energy in the mass of the house.

BB- The question remains, though, if it is an earth-powered house: Is the earth capable of carrying a house through a whole winter? That's where the solar heating comes into play, because the solar gain is needed to replenish the earth of its lost heat.

DD- I just want to clarify one thing. I wasn't at all advising that you not face your house south and not have south glass, I would never design a house without south glass.

<u>A-5 How much of the heat energy that comes in through the glass is getting down into the earth storage?</u>

NS- One way to look at this situation is that your daily heat storage and heat release is essentially all above waist level. The ground is a longer term proposition; the ground acts as a floor below which your temperature will not drop more than a couple of degrees, say in the greenhouse, because if it does, then the heat starts coming out of the ground. And you are replenishing that heat but you hope you don't lose it very often to a certain extent.

BH- You especially hope you don't lose it very often because, whether or not you think the earth is an infinite heat source, it doesn't move heat very fast at all. Whatever it's got down deep it won't be able to deliver until the month after next--we're watching the 6" deep temperature in the Burns house drop from 55° in October to 45° in February, and that definitely affects the performance of the house. I'd be worried about the late winter situation in Dennis Davey's house, where he's designing out the potential of the solar gain to recharge the earth battery.

B. ENERGY COLLECTION AND CONSERVATION

The Energy Collection and Conservation System is essentially the two shells of an air/envelope building, and its design variables are the amounts and positioning of glazing and insulation. It is Jim Berk's key insight that the same windows, with the same attendant heat gains or losses, can have remarkably different effects on the driving power of the convective loop, depending on where in the loop they are placed. For example, solar gain incident on the greenhouse floor gives the greatest push to the loop, for the air that rises there has the longest path to exercise its buoyancy. Conversely, glazing near the peak may be adding heat at a place where it has nowhere to rise to; this may even act as a drag on the loop flow, resulting in a condition of stagnation in which too much hot air is trapped at the peak and very little is making it down to the crawl space.

B-1 What does it mean to say that many of these houses have too much glass?

HH- The problem with overglazing, especially too much glass near the peak of the south face, is that it produces too high a temperature up top. This can lead to stagnation of the loop system, since there are limits to how fast the convective loop can really work. At this point you're pumping all that extra heat into a high heat loss area--because of the high delta-T and the very low R-value if it's glazed--and it's just going right back outside.

BH- Which may not be that bad during gain periods, but if the extra amount of glass isn't earning its keep then, it will still be eating heat during the other 2/3 of the day.

BB- The difficult thing about sizing glazing is that up to this point we really haven't gotten an accurate computer model of the thermal envelope house. So consequently we've been guessing.

DC- The envelope system seems to be fairly forgiving. Some houses are really overheating in the upper part of the envelope, hitting temperatures of 115° or 120°, but all in all they don't seem to be performing that much differently from, say, the Burns house, which only reaches 95° or so.

BB- This forgivingness is one of the problems about the envelope. It's such a forgiving system that we can't pinpoint what each part is contributing. If it wasn't so forgiving, we could say, "Oh, that doesn't work," but as it is it's going to require an extensive amount of study to zero in on what's really happening.

B-2 If the purpose of the loop is to add heat to the earth mass and perhaps the north wall, how can cooling it with north windows and skylights do any good?

BH- All the data on envelope houses I've seen to date indicates that a lot more heat is available in the upper parts of the envelope than is getting down into the long-term store. If 10 BTUs of that energy that's now being wasted can be spent in such a way as to bring 5 BTUs more into the store, it'll be well spent. Unless we go to fans, some heat loss high on the north side is the best positive way I've seen to encourage that warmed air to buck its nature and go down. Our data from the Burns house is a little sketchy, but the one of the five 8' bays that recorded the highest air flow was the one with the combination of less south glass and an 18 sq. ft. north window. That bay showed peak temperatures 10 to 15° below the greenhouse bay, yet its north plenum and crawl space readings were a couple degrees higher. (See Chapter 7 on Performance, p. 7.7).

B-3 Is there an optimum ratio of north-to-south glazing for augmenting the air flow?

--- Obviously you don't want to have as much glazing on the north side as you do on the south--that would be throwing away everything you just gained--but a smaller percentage of north glazing is desirable both for aesthetics and human comfort, for natural daylighting and for the possibility of cross-ventilation in the summer. These houses will certainly work without any north glass.

NS- You can tell to an extent a person's place in understanding the phenomena by looking at his design. The first design someone does has a greenhouse on the south and no windows on the north. The next one will likely have a high clerestory all across the north, because they want to induce some cooling there to make the air flow. They've discovered that the greenhouse doesn't work entirely alone; there has to be some cooling. Then second thoughts come in, and that is double glazed, and finally you get to the point where you will notice that the windows' penetration on the north is essentially independent, that it is put in there for people reasons, for architectural reasons, and it doesn't contribute much to the cooling.

B-4 What is the effective R-value of a window in the envelope?

BB- Heat loss and R-value are not the same thing. A membrane's heat loss is proportional to the temperature differential across that membrane divided by the R-value. Therefore, an envelope window which is experiencing 70° on the inside and 50° on the envelope side would have significantly less heat loss than a window directly exposed to the 10° ambient outside.

DC- I'm starting to feel that you could use the envelope in place of an insulating shutter on the east and west. In other words, you could put in an insulating shutter with a standard wall or you could make a double envelope right at the window section and use the tempering air from the basement below to maintain a temperature outside the inner window that would result in less heat loss.

—— If you put double glazing in each shell, you're quadruple glazing, and if you're quadruple glazing, don't bother to put the envelope on the east and west. The envelope keeps the inner surface of the outer shell warmer at night, hence it increases the heat loss. If you've got a sheet of glass and you proceed to couple it to your crawl space, the heat from the crawl space comes up and makes that glass warmer at night than it would be if it were not coupled to the crawl space. Insofar as that glass is warmer, the heat loss is greater.

--- That's heat loss that's coming up from the crawl space, though.

--- But that heat's got to come from somewhere.

--- Out of the infinite earth sink.

--- It's coming out of the earth sink, and it's not coming out of the inner house--that's the point. It's out of the buffering envelope. It's not infinite, no. Do you think it's infinite, Dennis?

DD-- Yes.

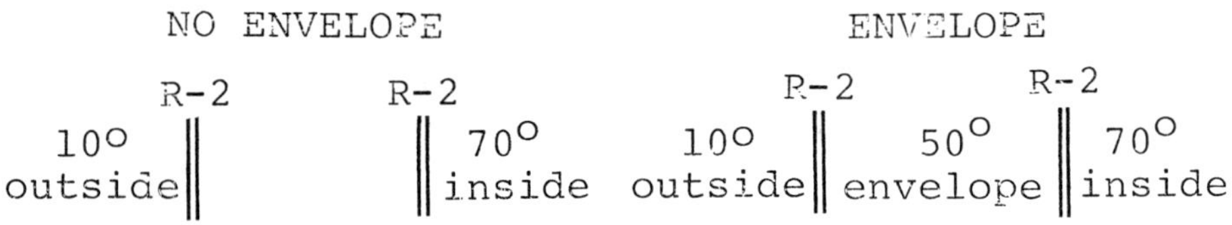

The simplified heat loss calculations according to the equation Heat Loss = $\Delta T/R$ for a unit area of surface illustrate some of the preceeding comments. In the upper example of simple quadruple glazing, Heat Loss = (70-10)/4 = 15. With the envelope in operation, the inner window's losses are reduced to (70-50)/2 = 10, whereas the outer window's losses are increased to (50-10)/2 = 20, with the difference being made up by a convective contribution from (ultimately) the earth in the crawl space. Thus, the amount of heat passing through the exterior glazing is actually greater in an enveloped window under severe thermal stress; however, the more moderate temperature outside the inner glazing will result both in less heat needing to be made up by the auxiliary heater and in less of a chill to the occupants' bodies from radiation to the cool glass. Working backwards, one might claim an effective R-value of 6 = ΔT/Heat Loss = (70-10)/10 for the enveloped window, as compared with R-4 for the regular quadruple glazing, but this value is largely symbolic, as it is determined by whatever temperature conditions exist at the time.

DB-- In a house that we've started building we've got the envelope on the south, east and west and not on the north, just because there's glass on the east and west and not on the north.

DC- I think any major glass area has to be protected by the envelope. You just can't have a lot of east and west glass and get the thing to work. The delta-T to the outside is just going to kill you.

--- Another way of saying it is that any glazing on the east or west ought to be improved to R-4 or better, with insulated shutters or whatever.

B-6 If a window is placed in a loop with its frame going through from the inner shell to the outer, how does it affect the airflow?

BH- Hopefully it is detailed so that it doesn't, since it is the glazing losses that are most dramatically reduced by the envelope. The air flow that counts is vertical here, so the sides of the window casing can be solid and cause no problems. Slatted or wire gratings top and bottom should permit the buffering effect as far as the window is concerned, but from the point of view of the air flow as a whole, it would be optimal if they were entirely unobstructed.

B-7 How important a factor is direct solar gain in achieving comfortable temperatures in the living space?

DD- If you're going to allow direct gain into the living space, you must size the internal mass to absorb the amount of heat that will be coming in; otherwise, the space will become uncomfortable.

HH- I find I get very little direct gain into my house from the sun; it's much more effective to open the sliding glass door for convective gain.

B-8 What about the inner south glass--how much should there be, and are there any special considerations?

--- Probably less is better, but you have to consider illumination.

--- I was coming up with 35 to 50% as the area of glass on the inner south wall as compared to the aperture of the greenhouse.

--- If you have 50% glass in the inner wall, you're going to have overheating in the summertime.

--- No, the sun's too high. It doesn't come in--it's bouncing off four sheets of glass.

HH- With the proper overhang, the inner wall is shaded.

--- I think it's purely an aesthetic point.

--- The glass should be double glazed, maybe with heavy drapes to protect it from heat loss to the envelope.

--- Draperies as they are normally installed will make us feel warmer because they shield our warm bodies from radiating heat to the cold glass but they will not reduce the heat loss of a room, since they create a negative chimney effect of cold air pouring out the bottom.

--- There's not really that much heat loss - with the greenhouse at 45-50° that's a Delta-T to the inside of 15-20°. I'm not saying you shouldn't use curtains, but you can get away without them.

B-9 Comments on the pros and cons of sloped glass?

--- It contributes to the summer shading problem, and overheating.

DC- You may want sloping glass for domestic water heating.

HH- Plants tend to like light coming from above, so the serious vegetable grower will want some overhead glass, especially for warmer weather growing. In northern New England the growing season is so short that it makes a lot of sense to have the warmer weather plants--tomatoes, green peppers, etc.--growing in the greenhouse right through the summer.

B-10 Does the angle of inclination of the sloped glass have much bearing on the heat gain in the greenhouse?

--- 20° plus your latitude is the rule for the angle of inclination.

HH- It turns out in my area where there is pretty consistent snow cover that the angle doesn't matter very much. You can get 10-20% more absorption through vertical glass with a bright snow field in front of you than if the glazing were perpendicular to the sun. From vertical to 50-55° it really doesn't matter very much. It is more dependent on the style of house and whether you want to deal with sloped glazing, which is harder to install and weather-seal.

B-11 What insulation values are being used, inside and out?

HH- I would like to share something from a conversation with Jim Berk. His suggestion is that not only do you want to keep the south glass low instead of high, but also to use a graded insulation if you can do it--little or even no insulation high in the north roof, increasing continuously as you go down the north wall. This is supposed to increase the drive of the loop on the north side by generating a taller column of cooler air.

--- You deliberately build a bald spot above the north plenum.

--- Seems like a logical place for a north-facing skylight on the high side of the roof.

--- In terms of specific R-values, Lee Butler recommends R-19 outside and R-13 inside for this part of the country--that's for a seasonal heating requirement of 7000 degree-days.

NS- On this business of insulation, insulate the outer shell as much as you can afford; with the inner shell there are other considerations that will determine the construction.

BH- I am aware that some West Coast designers are proposing reversing this emphasis. I presume that this relates to the idea of the earth as an effectively unlimited source of refrigerator-like heat. If that's true, then the envelope should be able to keep itself from freezing with even R-11 exterior walls, and more of your insulation should be devoted to protecting the inner house from the envelope. So here's an opportunity to demonstrate your faith in infinity. In either strategy I would more emphasize the difference between the walls that directly separate living space from the outside and any walls with envelope on one side. Just as you should shutter windows that are not buffered by the envelope, you should "super-insulate" direct walls.

B-12 Why insulate the inner shell at all?

HH- Speaking from experience, I asked myself that question about three years ago. According to my calculations the loop should run better if you didn't, and you want to get some heat into the inner house anyway, so I concluded "Why bother?" and I essentially ran the house that way the first winter. We did have some overheating problems, and we did also have a cool-wall effect at night because the circulating temperature, even though the wall was tight, was far from comfortable. Fortunately, I had made the envelope wide enough so that I could crawl into it and actually insulate it after the structure was built. If you intend to try it as I did, make sure you leave enough space, because it's not a very pleasant job.

C. ENERGY TRANSPORT

The Energy Transport System, essentially a duct from the top of the greenhouse to the crawl space which is so expanded as to serve the additional function of a buffering environment for the inner living space, is the defining feature of the envelope house. It is surely the least understood and most controversial in terms of the factors affecting its performance and its ultimate potential and liabilities.

C-1 Is air really going around the envelope in a simple loop?

NS- I think that if you watch the airflow over the course of a day in any one of these houses, you'll find that there are very many different patterns. You can have a single stream, two, three, or even five streams, all in the same air flow space, moving in different directions, and you'll get transitions from one condition to another as the day goes on. The five is a very transitory thing; the one is the major loop running during the early hours and the afternoon, and so on.

DD- Nothing horizontal--in my house there's no horizontal air movement. The idea is that all the air movement would be vertical. When I say movement, I'm talking about hundreds of different currents of air going up surfaces. I think it's the losses along the surfaces that cause all the air movement.

BH- One general point of agreement is that it's not so easy to get warm air to sink; thus, it's the gain cycle of the loop that requires our sensitivity and ingenuity in terms of positioning heat inputs and outflows, streamlining the airflow, and so on. Conversely, nearly everything works to get the heat up out of the ground. Dennis Davey has chosen to quit trying the hard part and to make the most out of earth-tempering effect; again, it all comes down to how much heat can you get out of the earth, and how fast?

C-2 What rates of flow have been observed?

HH- Well, I was part of the original measurement at the Burns House using the pitot tube, and I guess there's enough correlation between those results and what Al Converse recorded -- up to about 45 fpm there. Bob Mastin has told us that 70 fpm seems to be the maximum for his house, so it seems like we're getting something less than a foot per minute, down to a third of that, which is again what Jim Berk is saying. I talked to him earlier -- 22 fpm seems to be typical, and the kind of measurement you really want to get. If you get more than that, you've probably got too much construction at that point.

NS- The one thing that is critical about airflow is that the air move enough to remove the heat from the greenhouse to a place that is insulated for storage.

DC- The amount of heat you can move is limited to the heat capacity of air times the temperature drop around the loop times the volume of air being moved. It's the volume of air moving that counts, and it doesn't seem like any of these houses are running more than 2 cfm per sq. ft. of glazing. If that's true, why aren't they overheating more than they do?

NS- Because half of the heat goes right out of the glass, half of what's left goes into the structure, and one quarter circulates.

C-3 What dimensions are recommended for the width of the airflow spaces?

RS- We have this house near Oslo, Norway, with an interstitial gap of 2 inches--that's with forced convection.

DD- One house I've done had an 8" space and the second one had a 4" space. The second one hasn't been occupied, but during construction from my observations and from various temperatures taken at random I haven't seen any significant difference between the 8" space and the 4" space. In fact, I think the smaller theoretically might be the better because as you increase the width of the space, I believe you get a lot of internal turbulence in the space.

BH- It should be pointed out that Dennis Davey may be getting away with the smaller plenum because he's limited the amount of heat transfer he's trying to accomplish. Rather than trying to move a large excess of heat from a greenhouse down into the earth, he's simply trying to equalize the temperature of the small volume of air in his envelope with whatever temperature the earth can provide.

DH- I've tried to take the ASHRAE procedures outlined in the Handbook of Fundamentals and do air duct design dimensions, and I've come to the conclusion that in our kind of geometry the significant losses are at the expansions and contractions and at the corners--they're not in the straight sections of the ducts at all. The frictional losses are negligible compared to the momentum losses at 12". I then roughed out the calculation of 6" and it looks to me like the answer's going to come out about the same.

NS- The ratio stays relatively constant, regardless of the duct width, but the magnitude goes up as the square of the reduction in the separation.

HH- Berk says not to constrict the loop area at the connections between loop-legs to less than 90% of the minimum cross-sectional area away from the connections. The transition between the loop legs should be very smooth, since it appears that most of the loss in force driving the loop is in changing momentum where the air encounters directional changes.

C-4 Is it valuable to curve the corners of the ducts?

NS- The only data I have worked with extensively has been on the Burns house, and the only computations I have made are on a house similar to that house. In that case, the principle resistance to flow appears to be the dynamic resistance, which is most pronounced where the north plenum empties into the crawl space. It probably is a good idea to put turning vanes in there... However, you must remember that the flow is a square-law proposition, so it's rather insensitive. Not knowing what the actual flow resistance was to plug into my computer model, I ran a number of tests with different values and found that I could vary the flow restriction by a factor of four with almost no change in the temperatures that resulted from the computation... You have a temperature difference between the greenhouse and the north wall. That temperature difference results in a density difference, which gives you a pressure difference. That pressure difference then forces air around the loop, up to the point at which the pressure drop is equal to the pressure you had driving. That flow resistance seems to be primarily in the dynamic pressure drop; that is, if the air that's moving has momentum, and if it changes direction suddenly, as we do in most of these houses, you lose that momentum, you lose that energy... If you make the turns easier you get a higher rate of flow. With a higher rate of flow you get better heat transfer between the air and the surfaces. The better heat transfer means that the temperature differences are less, so there's less pressure to drive the air, and the flow slows down. It's a self-adjusting system, you see; it isn't very critical.

C-5 What about laminar flow?

HG- Turbulent flow mixes the air and has a higher heat loss; laminar flow between two walls would have much less heat loss to the outside, but it's very hard to maintain it in these types of spaces. We don't have a laminar flow going around the loops; they are turbulent.

JB- The distinction between laminar flow and turbulent flow is of little interest in convective loop design. It is not that laminar flow does not exist--it does. But it is limited to periods of low heat transfer, and little energy is involved. That's because, given the physical dimension of our container (the house loop) and the physical properties of our gas, then at any time when there is a significant main loop air stream and heat is being transported, any laminar condition gets quickly changed to turbulent flow.

C-6 How does opening the greenhouse to the living space affect the flow?

--- It doesn't really seem to have much effect at all. What it can do is siphon some of the heat off from the greenhouse and circulate it to the living space. In that sense it may tend to reduce the temperature of the greenhouse slightly, which would tend to reduce the flow slightly. But most of the flow is not going to the house itself.

HH- It makes very little difference if there's only one opening. If you open north and south, that's different. In our house we have a double sliding glass door on the north side, leading to a screened porch. Just open that north slider and open that sliding door onto the greenhouse, and you partially short-circuit the loop and get direct convective gain right into your house. At shoulder level the temperature is probably 80 or 85°; you get a real blast of warm air on a typical winter day.

D. ENERGY STORAGE

The Energy Storage System is twofold; heat that is absorbed by the structure of the building and excess heat that is stored in the house foundation and the upper 6" of the earth in the crawl space is used to moderate the envelope air at night and during sequences of sunless days. In addition, the seasonal build-up of heat in the first several feet of earth acts as a long-term reserve which slowly discharges over the winter, when withdrawals from the store exceed deposits.

D-1 How should you insulate the earth store?

JB- The area of the earth below the house should not be insulated (thus limiting the storage mass), but an insulation fence must be carried down the foundation stem to the footing level. In our past experience we have concluded that the earth mass beneath the building was not sufficient, and have reached out for more: moving the foundation insulation to the outside of the stem, and then moving it out even further to capture even more earth mass. Careful measurements will show, I am convinced, that this was unnecessary. The difficulty with the store of the earth is in the transport subsystem--not the volume of the earth.

DD- I always insulate from the footing right up to the bottom of the sill all around the outside of the foundation to keep that thermal mass that you've already paid for.

JF- Having plenty of insulation on the foundation cannot be overstressed. I think one of the reasons we experienced a drop in the earth temperature during the winter for the house in Virginia may have been that the foundation insulation is not heavy enough. We used 8" block in the foundation, filled with perlite, with 3/4" Dow Styrofoam on the outside; this went down to two feet, which is the frost line in Virginia, and then out three feet with 2½" beadboard.

BB- Concerning heat flow through the earth, my experiments indicated that the first inch of styrofoam is crucial, the next inch helps some, and three inches helps a bit more. Another observation was that one inch of insulation six inches below frozen ground is capable of keeping the ground beneath it from freezing, which raises some interesting questions about running footings four feet in the ground.

DC- We've heard of an instance where a foundation wall cracked under the pressure of frost-heaving earth just outside it; this was blamed on the fact that with an insulated foundation there's no heat coming from inside the building to keep the earth from freezing right next to the wall. This may be another argument for only running the insulation a foot or so below grade, then out laterally 4-8'. This also includes more earth mass in your insulated store. (See the University of Minnesota's book <u>Earth-Sheltered Housing Design</u> for a detailed assessment of this type of insulation.)

PP- Remember that it's the air you're trying to insulate yourself from, not the earth.

<u>D-2 Has anyone considered the use of insulation or vapor barriers underneath the earth store?</u>

DD- I'd say absolutely put no insulation underneath an earth mass and no vapor barrier. You can handle any moisture coming into the building, and eventually the earth mass is going to dry out.

HH- On the other hand, if you have flowing water under the footings, you should isolate yourself from that.

PP- If you would excavate about two feet of the building, place a vapor barrier in the foundation area, and then put the soil back on, you'd have an ample supply of earth to work as a heat reservoir and a controlled amount of moisture.

HH- Yes, I've got a vapor barrier, then two feet of back fill and concrete blocks as a storage mass over that.

<u>D-3 Is it worth the effort to put more mass or mass with a higher heat transfer capability in the crawl space?</u>

HH- Well, that's what I've done in our house--150,000 pounds of concrete block. A rather arduous chore, to say the least, especially in July. It appears it may have been worth it, but I can't really say for sure. What I think could be improved is the coupling to the mass that's already there; you're essentially letting warm air flow across the top of the crawl space or basement area and you really don't have that much surface area for heat transfer.

WM- In a current house we are putting water jugs under there with a catenary of a flexible material so that the air will be driven into that space between the jugs and there will be some extra heat transfer there.

DD- My experience tells me that it's not worth the effort and the money to add extra internal mass downstairs in the form of water storage or phase-change material, given the fact that has been borne out today that only 20% of the total energy hitting the glass is being delivered down there.

D-4 Would north wall storage help the loop drive?

BB- One of the questions that comes up when you consider those figures is that if only 20% makes it into the earth store, where's the other 80% going? And if that 80% is a heat loss that's driving the loop, couldn't we drive the loop by losing some of that heat into storage, say, at the top of the north wall? This may get expensive when you're considering continuing the foundation up the north wall, though this is what Don's done in the Burns house by berming up the back of the house to the point where the concrete wall comes to within 4' of the top.

NS- The Burns house is probably one of the best double envelope houses that's been built, and one of the reasons it's good is that north mass. That north mass extends fairly high, and it was giving off heat that was keeping the loop running. Before I got that mass in my model, it was a very wishy-washy thing.

HH- Berk is saying that you need to absorb your solar radiation low down on the south side and then lose or store some of it high on the north. I'm thinking of the phase-change salt bag we saw. Picture lining the envelope, just hanging these things right down the north wall. Storing heat up high there should have the same effect on the loop drive as a row of windows or grading the insulation so that it's less the higher you get--in either case it's just heat being taken out of the air that's moving by.

D-5 What are these "phase-change salt bags"? What are their advantages and limitations?

DB- Many substances store and release a quite considerable amount of heat when they change from one physical phase to another, e.g. from a liquid to a gas, like water when it evaporates or condenses, or from a solid to a liquid, like some of these salts. The attraction of such a material is that it can store this large amount of heat in a small volume and without raising its temperature very much. There have been technical problems in packaging the salts in such a way that they do not stratify over many phase changes and lose much of their effectiveness. The bags which are becoming available are designed to be used horizontally in ceilings, which is a limitation regarding Hank's idea.

D-6 Are there any problems with using a basement instead of a crawl space as the lower leg of the loop?

DC- In the Burns house where we have a partial basement, the mass temperatures of the concrete and the crawl space earth are similar. The records from other houses don't show any obvious differences in performance relating to whether they have crawl spaces or basement. Moist dirt and concrete have very similar surface conductances, conductances, and heat capacities. The only major difference might be the ability of the earth to give and take moisture, and it's pretty unclear how important that whole issue is. So basically I don't think it makes much difference how deep the space is, whether it's a crawl space or a basement, earth or concrete, because the warm air is flowing up under the floor joists above in either case, and heat is being radiated down from that surface to whatever is below.

D-7 What about adding mass to the greenhouse? A hot tub, for example?

--- It would increase your heat loss.

--- It would increase your comfort.

--- It would sure need a cover 90% of the time, or you would have a huge amount of condensation on the windows.

--- Why would it increase the heat loss? One of the reasons the double shell house works is that heat is absorbed in the greenhouse behind just a couple layers of glass; then it's removed to an area that is much better insulated, so that it escapes more slowly than it would through the glass. Therefore, if you're putting your store in the greenhouse, you're keeping the heat where the conductivity is high and hence the rate of heat loss is high.

--- The same argument would apply to a Trombe wall in the greenhouse.

DD- If you're going to use a hot tub in a greenhouse, keep in mind that if you do not put window insulation on the south glass, the temperature of that space is going to drop down into the 40's, and that's pretty cool to be taking baths in. If you do put insulation on the glass, then the hot tub will act as a heat storage for that room.

--- It seems to me that it couldn't hurt to have heat mass in the greenhouse. Maybe you are increasing the heat loss, but you're also increasing the heat gain.

NS- If the mass is such as to have a low surface temperature, this would increase your heat gain. I think you have to look at the whole twenty-four hours and, without running the calculation, my gut feeling is that any mass in the greenhouse will be a net heat loss. This is one of those things you could only settle quantitively.

BH- The way mass increases your heat gain is not, of course, by getting more energy to penetrate the glass but by absorbing more energy at a lower temperature so that heat loss directly out of the greenhouse is less. It soaks up heat until the air around it has cooled off, in the evening, for example, and then gives it back.

DB- For this reason, the effect of a brick or earth floor in the greenhouse might be to extend the gain cycle somewhat beyond sundown; spreading the gain cycle over a longer period would seem to make it more efficient.

D-8 What effect would you expect from increasing the mass in the inner living area?

DD- I believe you don't need to add mass to the interior of the living space as is customary with passive homes which rely more on direct gain. My experience with the envelope house is that the internal mass provided by the sheet-rock, furniture, and so forth, could be 20 to 30 thousand pounds, and I think that's more than sufficient to smooth out any irregularities in the heat supply.

--- If you haven't got direct gain on the mass, it's not worth bringing it into the house.

NS- When I was first running computer models without any internal mass, the temperature drop of the living area was 2 to 3 times as rapid as when I put in what I computed to be a reasonable amount of mass. What I computed turns out to be slightly less than is indicated by the measurements taken in the Burns house. It is a very definite moderating influence.

DB- It's worth considering the style of heat delivery of your auxiliary system. A wood stove is as sporadic as direct solar gain. If the phase-change bags had been available when we designed the Burns house, I would have put them in the ceiling of the inner house to smooth out the heat gain from the stove as well as the solar.

E. CONTROL SYSTEMS

The inner living area of an envelope house remains partially coupled to the external environment, though the extent of the swings of temperature will be greatly moderated by the buffering envelope with its earth connection. The controls of an envelope house are those systems of the house which the resident can use to fine-tune the gentle internal climate to her/his personal comfort. These may include sources of supplemental heat; vents and operable windows for cross-ventilation, possibly in conjunction with earth tubes or other air intakes for summer cooling; heat exchangers; and moveable insulation.

E-1 Where should a supplemental heat source be located?

HH- Berk claims that the best place to put supplemental heating is in the envelope rather than the inner house, and I'm afraid I can't go along with that. That's a rather indirect way to provide the extra squirt of heat that you need.

NS- If you look at the simple physics here, you have two shells, and there's only one heat source. You want a given temperature on the inside. If you put the heat into the inner shell, it takes half as much heat as if you put it between the shells.

MK- One thing that has been said about heavily insulated houses and envelope houses is that it really doesn't matter where you put the energy into the thing, it gets moved around the house. That should allow a lot of flexibility in the type of system you use. For instance, Gordon mentioned the tiny Paloma gas boiler--you could use radiant baseboard or whatever, but you could locate the heat supply close to the center and probably get the heat all around the house.

--- When you build these houses so tight and then try to carry out a combustion process inside them, it seems to me you're asking for difficulty.

--- My interpretation of Berk's thought was that if you have a combustion process, like one of these small, efficient gas boilers, that you should put the burner in the envelope.

--- With its fumes. Because that's where your infiltration is, and where your ventilation will be. Your heat losses because of the inefficiencies of the flue pipe and whatnot are going to drift around in the envelope, doing their bit for the whole thing; meanwhile, you pipe the heat by radiant baseboards or whatever into the house, where you need it.

E-2 Could one use a wood-burning cook stove regularly?

BB- We have a cook stove in one of our envelope houses. The owners light it up in the morning and cook breakfast on it, and the house is usually cool enough to absorb that heat very easily. But if you are cooking in the middle of the afternoon, there may be discomfort.

E-3 How about a masonry stove, or a fireplace?

BH- A masonry stove or Russian fireplace works on the principle of a rapid, hot burn in a massive fire chamber that releases the resulting heat slowly over a number of hours, after the fire has gone out.

--- You'll find that an envelope house usually requires a smaller amount of heat. If you had a smaller, lighter stove you would actually get more comfort out of it and be able to control it a lot more.

BB- If you're running a fireplace, it generally will draw much more air than can be made up by your normal infiltration, either into the inner house or into the envelope. That's going to give you a negative pressure in the house, which is going to suck air in from the outside, one way or another.

E-4 How would you supply the air for combustion?

BH- If it's a fireplace you're thinking of, you'd better rig up a way to intake air from the outside.

--- Jay Sheldon has an excellent article on supplying surplus air--for fireplaces, primarily--in the Sept.'79 issue of Solar Age magazine.

BB- If you are going to supply air to a wood stove, it's fine to supply it from the envelope rather than directly from the outside. There is generally enough infiltration into the envelope anyway, for as little burning as we're talking about.

BH- You've got to have some air change anyway, especially in the winter when you're buttoned up; if you don't have a heat exchanger, burning a little wood's a way to feel good about some infiltrating.

E-5 How do you prevent overheating of the envelope?

BH- The two basic passive venting strategies that I'm aware of are exemplified in the Burns and Beale houses. In the Burns and similar houses the envelope is vented directly to the outside either by manually operated hopper-type vents high on the south or north or by louvered gable vents in the east and west that have insulating trap-door covers on the inside. In the Beale house and some other designs there are screened gable vents to an un-insulated attic-space, among the trusses or above the collar ties, and a trap-door is opened from the top of the envelope into this space. The theory is that the suction created by the hot air charging out these vents can be used to pull cooler air from the outside into the house down below, either through some ground-level north windows or through an earth tube. By selectively opening and closing sliding doors and the like, that air can be channeled through either the inner house or the envelope. I say "theory" because it turns out that folks have run into many unforeseen complications in designing vents that work well under a variety of weather conditions. More on that in the following chapter (See page 6.5).

A slightly more active approach is Dennis Davey's use of wind-turbine ventilators, which help pull hot air out of an attic-space above the envelope; he uses a system of trap-doors similar to the second group of examples mentioned above. Another active-passive idea that's been tossed around is the use of photovoltaic-powered fans to assist in ventilation or forced circulation during periods of peak gain. So you get to fans. If, after reading all the cautions and advice you can get hold of, you're still queasy about how well your vents will work, think about where you'll be able to put a fan.

--- Maybe we should create cupolas, like old Thomas Jefferson, and use those cupolas for summer cooling.

--- A solar chimney cupola, with a dark surface on it and south glass--a superheated chimney which would literally drive the heat out and pull air in from a cooling tube.

<u>E-6 Is there a rule for determining the size of vents needed?</u>

DC- In chapter 21 of the ASHRAE <u>Handbook of Fundamentals</u> you can find the formula which describes the stack effect under natural convection; if you know how much air flow you need to pass through the vents, you can solve for the size of the vents. To determine the air flow you need, you compute the amount of heat coming in through your glazing and figure how much air it will take per minute to carry that heat away--here you add in the important design choice of how much hotter you're willing to let that air get before it goes out. When you solve all this for the vent area, it usually comes out to be 1/10 to 1/6 of the aperture of the greenhouse glass. Remember that you also need open windows, doors, or lower vents of a comparable area for the cooler air to come in.

<u>E-7 What sort of experience have people had with earth tubes?</u>

DD- We have two air conditioning tubes, one about 120' long and one about 80'. One bit of experience that I had is that I went outside one morning at about 6:30--it was about 5° outside--and I stuck my head down into the tube and felt a rush of warm air. It was really a highlight for the day. There's definitely energy being lost through that tube system directly into the atmosphere.

JF- I'm an architect and a contractor in Richmond, Virginia. A lot of my experience has been with the cooling aspect of the house; I assume you don't have too much of a problem with that up north. We've got two 80' tubes, 21" in diameter. We had a situation where the inlet for the cooling air was approximately 50' lower in elevation than the vent in the top of the roof--the tubes sloped down 15-20' along their length--and we had too much static head to lift that cooler air up to replace the hotter air going out through the vent. We were actually getting cold air moving out the tubes rather than in. My recommendation would be to make the tubes slope only enough to drain and no more than that if the site does not require it.

RM- I have the same diameter tube, 60' long, and it pulls fine, but it's almost level. It slopes down gently from the house but then it goes straight up in a 90^0 elbow, so the inlet is higher than the outlet in the house.

--- I would like to ask Bob Mastin if the earth tube is his only air inlet in the summer, when he's venting the greenhouse.

RM- Very rarely do I have that open alone.

--- If you open the outside windows, how do you get the suction to pull the air through?

RM- It still draws, though not as well.

DB- It seems to me that an air tube would be more effective if it had a 2" layer of styrofoam about 6" above the top of the tube. That way you would be disconnected from the surface extremes, both winter and summer.

NS- I installed some earth tubes twenty years ago, and as a result of those experiences am no longer using them.

E-8 How valuable or necessary is it to use (nighttime) moveable insulation?

DD- I think night insulation should go on every window, including the greenhouse glass. There's only 3 or 4 hours of good solar radiation a day; you should make every attempt to keep that in the house--it's just too precious.

BB- The one area that's crucial in terms of night time insulation is windows on the east and west. If possible they should be reduced, especially the west because of the overheating problem on summer afternoons; if you have them, get shutters for them. As far as insulating the greenhouse, it's a question of how much heat you're drawing out of the earth versus what you're replacing. If you're lowering the earth temperature to where the system isn't working effectively, then maybe we should consider night time insulation to reduce the heat loss out of the greenhouse.

JF- I'm not that sold on maintaining the greenhouse at comfortable levels--the whole idea is to let that fluctuate. The house I built in Richmond has glass everywhere, and the coldest the greenhouse has gotten is 41°. You're improving the performance, but it's not necessary to the functioning of the house.

GB- As a homeowner, not having to put on night time insulation is one of the big plusses of my house. As a gardener, I compensate by growing plants that can take 40° nights.

F. ADAPTING THE CONCEPT

In the final section of this chapter we look into some variations on the general envelope theme, including discussion of what kinds of spaces can be located in the envelope, different envelope configurations, the adaptability of the concept to larger structures, and the problem of retrofitting. We close the chapter with a look at the arguments for hybrid systems and a survey of where different designers see themselves going from here.

F-1 What sort of usable spaces can be put out in the loop?

JB- Devise all vertical loop-legs as usable or at least as accessible space, and all horizontal loop-legs as crawl space (or basement) or attic.

HH- Well, of course you can do that with greenhouses, entry foyers, and you can use the envelope as a hallway for internal traffic. Keep in mind that if you're going to put stairways and walkways out in the greenhouse or in the loop that, unless you use insulating shutters or other more elaborate techniques, it's going to be pretty cool out there a lot of the time--typically in the mid 40's before dawn on winter mornings. Another good thing to do with the greenhouse space is to incorporate a domestic hot water heater, either a pre-heat system or a panel collector like I've done. Berk recommends closets and storage space in the envelope, but you don't want to allow the owners or occupants of the building to carelessly block the loop itself. Arranging storage in the basement or attic space in such a way that you're not inhibiting air flow is pretty important. About closet doors--if you put your closets in the loop, then your closet doors are what's separating you in the inner house, from the loop which is either warmer or cooler, so whatever you're storing in that space is going to experience those temperature changes, and the doors themselves are going to be cold if it's 40° in the loop.

BH- It may be worth noting here that the north and south legs of the loop are not quite the same in comfort levels. The north envelope, while it won't have all the cheerful greenery, will experience only half the temperature variation of the sun-space.

F-2 Does anyone feel it's necessary to put in an airlock before going into the envelope? In other words, are we blasting too much cold air into the envelope by using it as an entryway?

--- The envelope is a high heat loss area, and when it's in the gain mode, you've usually got energy to burn. When it's in the heat loss mode, it's already reasonably cool, so it's a less sensitive area to deal with. It's ideal for an entryway.

F-3 What about a second-floor deck in the greenhouse?

HH- I live in a house that has a second floor balcony, and it's wonderful space when it has just the right conditions, but normally when the sun is out and bright, it's too hot to be out there. From what I've gathered about other houses, this second floor space is really not too usable because of the extreme temperatures.

F-4 Can a basement be made part of the living space? Or is there a problem with keeping the temperature up below grade?

DB- If you think of an insulated interior space at ground level contrasted with the same space eight feet deeper, it seems to me there would be no difference in heat loss. You're having the same interior heat, the same envelope temperatures, the same insulation between those two spaces. The difference is simply that there's no easy way to get heat down into the lower space. It's not that it loses heat faster; there's just no easy way for it to get heat.

--- The question of the cool basement is a matter of heat stratification--the higher the column, the more stratification in the living space, whether above or below grade.

BH- A basement can either be part of the envelope or, if you run a crawl space under it and provide an air flow, it can be part of the living space. In the former it would be suitable for storage or possibly a workshop, if your work is active. If you want to have an office down there, include it in the inner house and put your heat source down there; the other floors will take care of themselves.

F-5 Where should you put a garage?

JB- Do not attempt to put a garage in the loop, but parts of the garage structure can be part of the loop if the occupancy separation is maintained.

HH- It can be built onto the north side of the house. One of Bill Mead's houses has the garage across the whole north face. You have to make sure that whoever owns the house is not going to renovate and make that into a heated space, because in order to have the convective loop work one side of the north vertical envelope has to be an outside wall. If there's heated space on both sides, you won't get the driving force to make the loop work.

F-6 What's behind the idea of the house with an envelope on the east and west?

DC- That was my idea, and it came about because the house was designed for aesthetic reasons to have east and west glass. It had south glass, no north glass, and four windows each on the east and west. After observing the Beale house and realizing that there was a lot of conductance from the inner house to the outside through the east and west windows, it just seemed that the biggest losses were out those windows, and that's where the envelope belongs, where there's a glazed area. As we've seen, the air flows every-which-way; it's hard to stop these convective loops, and I think there will be no problem in getting the air to flow wherever you put the envelope.

F-7 Are there any advantages to building a double wall on all sides of the house?

--- The question comes to mind: Do you get a more consistent flow if you have heating on one side and cooling on the other? Having the envelope on all six sides, with heat on one side and partial cooling on the two sides and the back, would this interfere with the convective flow?

--- You'd be gaining heat on the east and south sides for half the day and on the south and west sides for the other half. And there would be cooling on the other two, so I think it would work all right.

--- The air flows are so mixed up anyway, you couldn't get them any more mixed up.

DD- In both the houses I showed yesterday I constructed double walls, both east and west, but I did not connect them with the ceiling envelope. They were connected, however, with the crawl space, but I stopped the flow at the top. I'm sure it gives you a better insulated building.

F-8 How does the height of the building affect performance?

--- Well, the higher the loop, the more the pressure head can develop on either side, so, it makes some sense to build a taller house. As far as performance is concerned, it clearly isn't necessary. The Burns house seems to be working as well as or better than all the other houses that have been built, and it's only a one-story house.

--- What's becoming clearer in my head is that the heat's up in the attic and the storage is in the basement, and the closer we can bring those together the greater the heat transfer. Once you start building three levels, a basement, first and second floors, you get a lot of stratification in the envelope; you get a lot of stratification in the interior of the house, and consequently discomfort on the lower levels. There are two ways to bring that heat and mass together: one is to bring the mass to the heat, which seems to imply bringing the mass and probably the foundation up the north wall, which was done in the Burns house to some degree; the other option would be to bring the heat down from the attic through a duct with a fan.

--- Or improve the loop's effectiveness in transporting the heat down.

F-9 What are the possibilities for larger, commercial envelope buildings?

BH- Malcolm Lillywhite has designed a five-story convective loop apartment building which is completed and seems to be working. He's reported to have said that it was more cost-effective to use the loop in an apartment building than in a single dwelling, which makes sense. The usual advantages of scale are compounded in an envelope structure in that the ratio of envelope volume to living space goes down as the size of the building increases, as does the percentage of building materials which is "extra"--that is, required to enclose the loop. Also, the internal gains in a larger building become really significant; the cooling load might be the major problem.

DB- The estimated fuel consumption for the Sant Bani School that we designed, but that has not been built, was between one and two cords of wood a year.

BB- In our township a new elementary school was being proposed, and I went to try and discuss various considerations about it with the architectural firm. It's really a whole different ball game when you get into a large scale building. The lighting load is usually larger than the heating load. Then the building is overheating in spots and underheating in other spots. You take an elementary school--they've got 2000 students huffing and puffing, which is a tremendous gain during the day and not there at night. So it's a much trickier thing to design.

F-10 What are the potentials and/or problems of retrofitting envelopes onto existing structures?

WM- We've been looking for a way to retrofit our farmhouse, which is luckily facing due south and has a rather long east-west axis. We figured out the south greenhouse and the roof over the top, and the problem seemed to be, once you're down on the north side, how to get back in again. Assuming your dimensions are fairly minimal, which ours are, and you don't want to spend a lot of money moving the cabinets and that sort of thing, you really have to build a new outside wall, which is fine, because that's the one that has to be built really tight against infiltration. The real problem is getting an air return back through the foundation wall into the basement, so I'm three quarters of the way there, stopped at the north basement wall.

BH- So clear access to the earth under the floor is something to look for. I suspect it might be the case that other areas of the country besides New England might have more homes built on pilings with air already flowing underneath--mobile homes would be another example.

DC- Or if you've got a large, old house and you can overcome the vapor barrier problems and all that, you can afford to lose some inner space by putting the wall on the inside.

--- If you could make an envelope inside the outside walls, would it be as effective to insulate the inside of the foundation walls?

HH- If you can make thermal contact with the mass that you do have, then you're not losing too much by not including the foundation wall, unless you get into a situation like the Burns house where that north wall extends up above floor level, in which case it could become an important part of the heat storage system. In Berk's paper he recommends that you insulate the inside of the foundation wall, but keep in mind that he's in California.

BB- Often your foundations may extend two feet above the ground or so, and if you could lay your styrofoam on the existing ground or possibly dig it down a little, then continue up the outside of the wall and secure it with dirt, you would be berming the house up higher and insulating it at the same time.

BB- A project that won an award recently in a passive design competition in California was for a conversion of a house trailer into an envelope house. I think it would be a fascinating project to take on. I think any trailer owner would be pleased to have a greenhouse along the whole south side of his house, and the air traveling over the top would give it a little roof to make it not look like a trailer. It would definitely change the heating characteristics.

F-11 Has anyone experimented with forced convection in an envelope house?

HH- It's obvious that forced convection will move a lot more heat than even the most ideal natural convection loop. I've tried our house one season with and one season without, and I know we had a lot more heat transfer with the forced convection. With forced convection, you're allowed a far more favorable situation for transferring that heat to any kind of thermal mass, whatever and wherever it is, and you can certainly devise schemes to blow warm air through the crawl space in ways that would make higher contact with the thermal mass itself and thereby transfer more heat. If you tried to do that with natural convection, you're likely to reduce the airflow rate. I have some interesting comparative data on the two houses that I've built other than my own. One house is gravity convected and the other is forced; both have simple basements for storage, and they're within 4 miles of each other. The Peterborough house, in which we installed a fan system several weeks ago, started off with an average minimum basement temperature about 3° lower than the other house; it's gradually come up to where it's now running about 6-8° warmer. The maximum temperatures in the envelope space, first or second floor of the greenhouse, are much better controlled in the house with forced convection, typically running 15° or so cooler than the house without forced convection. The occupants are pretty happy with it--they'd like me to put in quieter fans, but I think there's a lot to be said for getting into some air handling, in spite of the fact that people are still trying to do totally passive solar.

F-12 If you go to fans, why not use a centralized duct instead the envelope?

--- The volume of air being blown through a duct would likely be comparable to the volume that flows naturally around the larger envelope, so what's the difference?

NS- In your double envelope design, that air is flowing over very large (surface) areas, and those areas have a high thermal capacity; so it's no surprise it loses most of its heat by the time it hits the crawl space. In the typical fan installation, the area that the air is in contact with as it goes down underneath is 1/10 as much and is specifically built to have a low thermal capacity. So I would advise any client not to use a double envelope house; a duct is simpler, smaller, etc.

DD- As far as I can see, the reason for using a double wall building is that you want to envelop the inner walls with air that has been moderated in temperature. If you use centralized ducts, you will be defeating that purpose. You will then have the inner wall of the building in direct contact with the outside, which is exactly what you don't want to do in a double wall building.

HH- It seems to me that it can be very useful to have some sort of forced convection aiding in the storage of heat, while still allowing the building to adjust itself to the heat demand mode, whether it's a cloudy day or at night. The envelope is a nice passive way to accomplish that goal, and if you decide to go that way, then the envelope is already there. It's a very nice low pressure duct; why not just design a fan system into it? If you're looking at moving a lot of air more directly through a smaller aperture duct, you're talking about a higher head loss in the duct itself, which means significantly more power to run the fan. Now I'm talking about circulating fans that use a couple hundred watts of power; with the pressure head so low, it's just a simple air circulator rather than a high pressure blower.

BB- I'm not saying do away with the envelope, but the crawl space is the essential storage area, so why not get your heat gain down to the storage immediately and get as much good as possible out of it before you release it to the rest of the house?

BH- Actually our figures show that 2/3 of our daily heat storage is in the structure of the house--that's the walls of the envelope--and are we sure that's bad? It's right there, cutting down heat loss from the inside as directly as possible.

HH- If you're charging the store from the south to the north with ducted heat--think of our house with the concrete blocks--this means that the south side is going to be warmer than the north in terms of temperature stratification through the mass. Now at night, you've got a situation where under normal circumstances the air would want to flow from south to north through the storage, and then up and around the loop, down through the greenhouse and back into the mass. In this case, you've got the temperature profile reversed, so it's going to be working against the passive flow pattern. In general, and this goes for rock beds too, you want to charge your storage in one direction and discharge in the other direction, so that the air coming out contacts the warmest part of the storage before it's delivered to the load.

BB- In most envelope configurations, it would be very difficult to set up a whole row of fans that closes off the envelope and then pushes the air around. That brings you to the biggest problem with using the envelope as a duct, which is that the system has to run passively most of the time; therefore, it has to be left unimpeded. It's much simpler to use just one small duct that runs from top to bottom.

HH- That small duct is still taking up room in the house. The system I've set up in my house is very simple, and works excellently. It's in the upper corner of the basement on the north side, just some very simple circulating fans tied up with a polyethylene flapper-damper. When the fans come on, they pull the damper right into position so that all the air that the fans want to move has to come from higher in the envelope--in other words, they can't be short-circuiting themselves, blowing air right around themselves. When the fans shut off, the flap drops open and allows a natural gravity convection loop to occur at night or on cloudy days. It's pretty simple. (See page 4.9).

<u>F-13 In what directions do the individual designers see themselves moving in their next projects?</u>

DD- I am working from the idea that the envelope's average temperature is derived more from the energy of the earth than from the solar gain. The solar, as I see it, just gives daily impulses, and I don't really care if this solar energy gets to the earth mass or not, because we're relying on the constant outflow of heat from the earth to balance the temperature of the envelope. A 4" air space seems quite sufficient to do that; the thousands of local convection currents temper the space within each wall, with relatively little air movement around the loop as a whole. Where I'm going from here--the houses are going down in glass area to about 8-10% of the floor area. I'm not using doubled windows; it's just going to be one thermopane window connecting the inside to the outside, and all the windows will have thermal shutters on them.

DC- I am enthusiastic about the idea of using a fan during the peak gain period. I like the idea of moving the heat to the storage; I think it's a simpler and cheaper strategy than moving the storage to the heat. The house I'm now designing is a partial envelope that forces warmed air through interior masses and into the basement.

HH- I think we need not be too tied to things that have been done in the past. The ideal is a totally passive system, if that in fact will work, but I'm open to some sort of modification, some other way to move air around more efficiently. The combined system that I think holds a lot of promise is one where the envelope system is supplemented with some daytime forced convection to handle the high heat in the peak; this would allow large amounts of south glazing. Within the purely passive envelope idea, I think the trend is toward optimizing the factors that influence air flow, which seems to mean reducing the amount of glazing, refining the placement and quantity of insulation, and storing the heat in the most convenient and economical way. I think that in five or even two years from now we are going to be a lot further ahead than we are right now.

DB- I believe a 100% purely passive solar house is attainable through refining the basic envelope system, with higher earth storage and envelope temperatures and reduced losses through the glazing. One idea toward the former is a batch water heater which is emptied into the crawl space on sunny days. For reducing losses, I prefer less glass and triple glazing to shutters. By improving the semi-tempered environment of the envelope we can make it both more useful as living space and reduce the supplemental heating load to practically nothing.

6. Discussion on Construction & Special Concerns by Bob Henninge

This chapter will be concerned with questions relating to the actual construction and maintenance of an envelope building. The subjects that will be addressed are these: Site Considerations, Framing Systems and Details, Insulation and Vapor Barriers, and Fire Protection, followed by a discussion of some specific concerns about the Interior Environment and some thoughts about Costs.

G. SITE CONSIDERATIONS

BH- The prevailing solar wisdom is that a house should be oriented to face within 20° of true or solar south, possibly favoring east-of-south to speed up the gain in the morning and decrease the potential for afternoon overheating. These considerations apply in general to envelope houses, though it is less critical, since the solar gain is not used to heat the house directly.

G-2 What is the optimal shape for an envelope house?

BH- Because the envelope is at least as much an insulating system as a heating system, the square floor plan, which has less surface area, would seem quite favorable, as well as the rectangle elongated on the east-west axis, which would maximize solar gain. A north-south rectangle would be less desirable from both these standpoints, with an additional possible problem in that the envelope air would have to travel long horizontal loop-legs in which the pressure head would become attenuated with no relevant gains or losses to pump it up. Another case for forced convection, I suppose, if you have to orient that way.

G-3 Can you build an envelope house directly on ledge?

--- In the second house we built we had to blast out 8' of ledge to put the house in; we only excavated down to the footings. I see a possible advantage in that you've increased your thermal mass by that rock underneath and you've eliminated moisture underneath the house, presuming you've put in footing drains.

--- I built my house on solid ledge, and it's working real well. The only problem I have is that it's a two-story house and the bottom story is part of the air flow and isn't heated as well as the top.

BH- If the ledge you're building on is exposed to the surface anywhere near the house, you've got a large hole in your foundation insulation, and I would expect problems accordingly. Some creative insulating and backfill could help.

G-4 What problems are posed by ground water under the house?

JB- Static pockets of water improve the thermal store, but moving water near the surface adds to the losses. When the nearest aquifer is more than the sum of twice the depth of the foundation insulation plus half the least-width of the house, then the aquifer is of no concern, as all of the losses are then through the earth to the surrounding air.

HH- The site that we had to contend with showed actual flowing water when we were doing the excavation. What we did was to try to obtain as much thermal mass above the footing as we could, and in fact we did insulate underneath the floor slab in order not to have heat sucked away by the flowing water.

DB- At the Burns house we thought interceptor drains might be required to prevent water moving down this south slope from taking the heat away from under the building. Eventually we decided not to and instead just put footing drains around the structure. If perforated pipes are properly installed they're self-cleaning--I've been amazed how many people don't even recognize that the holes go at the bottom--it's important to cover the pipes with a foot or more of clean gravel or stone so that soil particles are removed from the water on its way to the drain. (See photo page 2.2).

H. FRAMING MATERIALS AND DETAILS

One of the original attractions of the envelope house was that it was a workable passive-solar design that could be built entirely with commonly available building materials and according to standard construction practices. With no roof ponds, rock beds, Trombe walls, and the like, let alone complex collector systems, it held out the promise that an innovative designer could conjure up some drawings, call up a friendly contractor, and be living in the 21st century in a few months. This was perhaps slightly optimistic. While the materials and techniques required are common, the consciousness with which they must be applied is not. Tales of woe are too commonly heard from solar enthusiasts and architects alike concerning the contractor's neglect or skimping on certain details that "don't show" in the finished house but can be absolutely crucial to its performance. Until these awarenesses are more widespread it behooves any designer to be utterly specific with his builder about details and practices and to communicate to him as fully as possible an appreciation for the principles at work in a convective loop building. This same need for care and thoroughness could make the envelope house ideal for the studied but less experienced owner-builder.

H-1 Is standard stud-frame construction suitable for envelope houses?

JB- Use batt and stud construction on all outer, inner, and direct membranes for maximum utilization of the thermal store and for the most economical construction.

BH- The envelope house is a light house inside a massive house. The idea is that the outer, massive house supplies a moderated environment for the lighter, inner house, which will respond rapidly to the resident's opening a door to the greenhouse, burning a few pounds of wood, or operating other control systems to adjust the interior comfort level. This makes stud construction perfectly suitable for the inner shell and, since the house's mass is concentrated in the earth underneath, the outer shell may be similarly light. The main complication that comes from the use of stud framing is that, since the floor and ceilings are either set upon (platform) or hung upon (balloon) the walls, the connection between the joists and the walls results in multiple penetrations of the vapor barrier and leaves a lot of exposed wood in the envelope. This is considered less than optimal from a fire prevention standpoint. A similar caution has been raised against sending the air flow through an attic full of trusses. (See the structural details that were used in the Burns house--chapter 2--for an example of a better system.)

H-2 What about post-and-beam construction?

BH- The common use of large thermopane panels on the south face leads you to a post-and-beam structure there, and the considerations mentioned above influenced us (in the Burns house) to choose a system of large supporting members (laminated box-beams and trusses) 8' apart, which is essentially a post-and-beam approach. A combination seems to work well.

DC- There can be vapor barrier complications with post-and-beam if the insulated wall sections are set between the framing members.

H-3 Has anyone used stressed-skin sandwich panels?

BH- Large panels (4'x8', 4'x16' and possibly longer) made of two surfaces (aspenite, plywood paneling, drywall, etc.) which are bonded to a core of foam insulation are now becoming available. The panels are extremely rigid and form a tight and sturdy wall when placed on posts 4' apart. Such panels in even larger sizes (8'x40') have been in use in Scandinavia. (See photos page 9.2)

DC- To use these panels efficiently it is important to consider their special strengths and limitations (rigidity, modularity, difficulty in handling) early in the design process. They are capital-intensive, but might save considerable labor time for a contractor who used them extensively. Probably less useful for the owner-builder, unless you also laminate them up yourself.

H-4 What difference is there in the treatment of the inner and outer walls?

BH- There are differing opinions as to which wall should be more heavily insulated; in California it's apparently common to use R-11 2x4 walls for each. It is clear that the outer wall will be subjected to considerably greater thermal stress and vapor pressure, and it's generally agreed that that wall should have the vapor barrier on its inner surface. Any walls that directly separate the living space from the outside deserve added insulation and extra attention to preserving the integrity of the vapor barrier, since they have actual, live wind and cold on the other side.

DB- A section of the wall we used on the east and north of the Burns house has 2x4 vertical studs on the inside, 2x4's outside the studs nailed on edge horizontally to support the vertical boarding, ripped 2x8's for plates and at window openings, with crossed layers of 3½" fiberglass.

BH- Another option, if you're using horizontal siding, is a version of the offset double-stud wall people have used for sound-proofing. This further reduces heat transmission through the wood. You can use the diagonal bracing that's let in to one set of the studs to tie the whole thing together.

BB- A similar system I'm using on my houses this spring: 2x4 walls instead of 2x6, putting Thermoply or some sort of tight sheathing which is a vapor barrier on the inside up against that wall after it's insulated, then running 2x2's horizontally on the inside of that. Now all the wiring goes between the 2x2's and the drywall gets nailed to them, too. The Thermoply is reflective and totally eliminates infiltration; heat that passes through the drywall bounces back because there's an air space with a radiation barrier, and there's a tight vapor barrier that's not being punctured every time someone hangs a picture or misses a stud with a drywall nail or puts in an electrical outlet.

<u>H-5 What is the construction sequence for the various layers of the two membranes?</u>

BB- Try to think out all the elements of that double wall and roof arrangement so that you can optimize the times for constructing them. If you are going to frame two separate walls, build the outside one first and let the inner one go until later.

DB- You don't want to put up insulation or sheetrock until you've got a roof to protect them.

BB- Once the roof and windows are in place, you can insulate, vapor barrier, and drywall the outer wall, then frame up the inner wall, adding the drywall to the envelope side just before you stand it up.

<u>H-6 How did you fasten the sheetrock on the upper surface of the ceiling framing?</u>

BH- We just slid it up there as we were framing and tacked it.

<u>H-7 What sort of surfaces do you want on the inside of the plenum (air flow space)?</u>

JB- Keep the ceiling surface of the crawl space non-reflective, dull and made of common building materials--kraft paper, plywood, gypsum board. Up to half the heat transferred from the air to the earth can be by way of radiation from the floor undersurface above. Elsewhere, the loop facing surfaces should be reflective, if possible, but it is not vital. Use aluminum foil-faced insulation with the foil faces on the loop side.

BB- In my first house I used reflective foil inside the envelope, feeling that the heat loss from the house would be reflected back. Now, at this conference there's a lot of talk that it should actually be drywall, which picks up the heat and later radiates it out.

DB- I decided to put sheetrock on all the envelope surfaces because I didn't like the flammability of the polyethylene vapor barrier and wanted to hold the insulation permanently in place. Looking back, I'm dissatisfied with the amount of heat that never makes it to the crawl space because it's absorbed in all that gypsum, and I would like to find a less massive but non-flammable surface covering--possibly a sheet metal.

H-8 What do you do with windows through the envelope?

BH- A common detail in a number of early designs seems to be to bridge from the inner window to the outer with a wooden lattice or grille-work that allows at least some air to circulate between the windows and visually keeps you from looking into the envelope. It's my impression that this strategy allows enough flow to temper the air between the panes but blocks the window from making a significant contribution to the flow in the loop as a whole.

HH- Here's a north window detail that shows the Andersen Permashield unit on the outside. No grille-work on the sides because the air flow is expected to be up and down. What I've done on the top and bottom is to omit any grille-work at all for wintertime operation. There's a plexiglass sliding pocket window that we site-built, which just slides into a pocket in the wall on the west side of the window, so that in the summertime you have just the single window to contend with, and at that point if you wanted to use that space for a flower pot or something you can put in the grille. (See diagram and photo, pages 4.7 and 4.8).

DB- At the Burns house we have a six foot window in the outside wall and a four foot window in the inside wall; this avoids the cave-like impression of a tunnel through thick walls. We painted the outside of the inner wall black so that when you look in from outside you don't see the inner wall.

BH- Again, it's a legitimate choice to have a single thermopane window, like you probably will have on the east and west, and just get some good insulating shutters for it.

H-9 What types of vent designs have been tried?

HH- Venting can be a problem. At the February 23rd conference there was an example of some north-facing ridge vents that were improperly sized. They were using metal-clad door units which turned out to be too wide in that situation. Hot air was going out the top of the vent and cooler air from the roof was going back in the bottom of the vent and short-circuiting the draw so that the greenhouse was overheating. They had to block off all but 8" of the opening before it would pull right, but by then what had happened to your vent-size calculation? You live with a hotter greenhouse than you'd like.

JF- We're now using a power exhaust system in the attic.

--- I've talked to several people who thought they would ventilate out the south side, and they found the radiation off the roof in the summer was so great that it effectively blocked the ventilation out. They were using a south-facing clerestory that got too much heat off their roof, and it negated the ventilation.

BB- It's a very tricky thing to design a ridge vent that will vent well enough to do passive cooling. If you run your vents up the south side, the wind can come up from the southwest and pressurize the vents so that the air won't come out. Someone had a clerestory vent facing north, which seemed optimal; however, they had a black shingled roof, and the heat coming off that roof also was so intense that it nullified the hot air inside the greenhouse trying to come out and actually forced hot air back into the vent. There's also the difficulty of sealing dampers on ridge vents and keeping water and moisture out. It's something that really needs a lot of consideration.

HH- In general, the venting arrangement should consider the prevailing wind direction and should be oriented such that the wind isn't blowing into the vents most of the time.

BH- It's definitely a complex question, and we've given you some things to ponder on, unless you're willing to head straight for the electrical solution. Ridge vents appear to be very tricky, because the roof surface just below the vent opening will be creating its own air currents. Any vent that opens in only one direction will be vulnerable to wind interference, and the direction in which it opens had better be checked out with the Weather Bureau. Hank seems to have no complaints about his gable vents--he's used them with insulating covers, where they vent the envelope directly, and in an uninsulated attic space, into which he can dump his excess heat through a pulley-operated trap-door. Gable vents are bi-directional. Following that idea further, we arrive at the time-honored cupola, which has vents in four or even eight directions. Wind turbines, photovoltaic fans, and creating a solar chimney to help push the air out of your vents, are some innovative ideas that have been suggested. What's yours?

I. INSULATION AND VAPOR BARRIERS

I-1 What about improved insulations?

DH- I have an axe to grind on that one. Still air is a good insulator. You don't hear it referred to as such but I think if you do the arithmetic, if you convert the conductivity of dry air to an R-factor, it's about 5.5. Anybody who tries to sell you an insulation foam material who claims an R per inch of greater than 5.5 is giving you the value at the instant the foam is made; it's only correct for a short period of time. They use a gas to create that foam, and the gas may have a better insulating value than air does, but unfortunately that gas will not stay there--it will diffuse out through the polymer and be replaced by air. Therefore it is my sincere belief that within a number of years, maybe five to ten, eventually there is no foam that can possibly be better than R-5.5 per inch, because that's air, and there isn't any better material around.

DH- If expense were no problem and you were to take urethane and hermetically seal it in aluminum foil, 5 or 6 mils, not only on the lateral surface but around the edges, if you literally encased it in aluminum foil, I think you could maintain that value of R of 6-8 per inch.

--- The phenomenon of Freon leakage can be largely prevented by spraying a high density skin on both sides of the urethane foam--I believed that's what Advanced Cooler Products has done--or by encasing it in metal or foil or plastic as in insulating space craft.

BH- The converse to Doug's comment that dry air is about the best common insulator around is that water is about the best common heat conductor around. Fiberglass and cellulose have their insulating value drastically reduced when even a relatively small amount of water vapor replaces that air, hence the growing insistence on tight vapor barriers as a part of the insulating strategy.

I-2 Has anyone investigated this new ceramic insulation?

BB- It's kind of wild. They claim that they encase the Perlite in silicone by a heat and pressure process which encapsulates the air spaces and gives it some new resonating effect like R-70 for 3" thickness. Which is certainly interesting--a vacuum doesn't give you that. I'm doing some tests on it and it seems to be performing as good or a little bit better than 6" of fiberglass--that is compared to 3" of the ceramic. It is, however, three times more expensive. It can only be used in ceiling applications. According to their instructions, it has to have air movement over the top of it, otherwise it drops from an R-25 per inch down to an R-2.5 per inch. I couldn't get them to explain how that could happen. I've been told that it works amazingly at very high temperatures, and that's why it's been used in space vehicles. I don't know--in a low temperature application I think it's closer to Perlite in value.

DB- My judgment is to be cautious about it.

--- All the insulations have their good and bad points; there's no miracle product that will make up for careless installation.

I-3 What parts of the house deserve special attention, as far as insulating is concerned?

JA- We scanned our house with an infrared heat-viewer, and that's when the defects in workmanship showed up. It showed black, which means a lot of heat loss, wherever there were interior stud walls meeting exterior walls and a corner had been boxed out for the sheetrock; there were black lines all along the ceiling connections and all around the walls, at the plates, and especially at the corners. Also in the attic where we insulated between the collar ties; they weren't exactly 2' on center, so there's extra loss there.

DB- My observation is that the concern of carpenters and builders and contractors is not necessarily to accomplish the best possible construction but to build a house quickly and at moderate cost. Corners are quite commonly not insulated, and you're smart if you have a bag of insulation scraps handy when the framing is done so that you can go around and fill all those trouble spots while they're visible, before anyone is thinking of insulating.

HH- A trick that's useful with fiberglass batts in a ceiling is to take a bag of cellulose and spread it over the top of the fiberglass; it works very well to fill in all the gaps and decrease the loss from infiltration as well.

DB- Another point that builders often neglect is the need to allow ventilation of the insulation just under the roof sheathing, to permit moisture to evaporate from the insulation and to keep the roof surface cool in winter so snow won't melt and re-freeze at the eaves.

I-4 What kind of protection is used for exterior foundation insulation?

BH- Foundation insulation is vulnerable to deterioration caused by ultraviolet radiation above the earth surface, moisture below the surface, and insect and rodent damage in either place. Of the three common types of rigid foam, only extruded polystyrene (Dow Styrofoam) has a closed-cell structure and will not absorb moisture. It can reasonably be used unprotected below grade. It seems a good idea to protect urethane or beadboard (expanded polystyrene) with some cementitious coating or a layer of polyethylene vapor barrier.

DD- One means that we used was to cover it with asbestos board above grade; it goes on very quickly, and we carried the siding down about an inch past the top.

DB- At the Burns House we used Dryvit, a resinous, stucco-like material that goes on in two coats around a reinforcing cloth mesh. Standard Drywall's Acryl-60, mixed with Portland cement plaster, and applied with a similar mesh, also works well and costs less.

I-5 What is the purpose of a vapor barrier?

BB- When air is warmer, it's capable of holding more moisture. As the air cools, its ability to hold moisture decreases until it reaches the point where water in the air will condense out--that's the dew point. Therefore, no matter how absolute humidity there is, the only time you see condensation is when the air temperature has dropped to the point where it can no longer hold that much water, and the excess condenses out. That will only happen in a house where there is a large temperature difference between the air and a surface. That's likely to occur on any glass surfaces that are exposed to the outside. It could also happen in the insulation, and that's the place we can't tolerate it, and that's why we talk about vapor barriers so much. If the high moisture levels that are present in the envelope find their way through a breach in the vapor barrier, they may get into the insulation, which significantly reduces its effectiveness, and when that moisture contacts the cold sheathing it may freeze. Later on it will drift down to the plates, and you can have water on the floor or rotting wood.

--- The other thing that's important besides a tight vapor barrier is allowing the outside skin to breathe.

BB- There is an architectural rule that says your outer skin should breathe five times more than the inner skin. Now drywall has a perm rating of 50, plywood has a perm rating of only 1 or 2 because of the glue layers, and Dow Styrofoam is less than 1. All these builders are slapping Dowboard on the outside, taping the joints, sealing it up, and putting drywall on the inside with kraft paper, which is a lousy vapor barrier. Fifty times the transmission of moisture through the inner wall as opposed to the outer. It's just got to be condensing when it gets to the outside wall. The only thing that's probably saving them is that the Styrofoam insulates well enough to maintain a moderate temperature in that cavity. But it's bound to be wetting the fiberglass, and fiberglass loses tremendous amounts of its R-value as soon as it's wet.

DH- My engineering approach to things is always to sketch out the extreme conditions first, to find out how big the ball field is that we're playing in. I decided in this case that the people who probably knew the most about heat-tight housing lived in Finland. I wrote to a Finnish friend and he eventually sent me a stack of papers from Swedish building research. The essence of their comments was that there was no new discovery in insulation. They spent as much time talking about how to construct vapor barriers as they did about how to insulate. Shortly after that, someone suggested I send $4 to Ed McGrath up in Alaska and he very kindly responded with a little book "How to Build a Super-Insulated House"--37 pages. (Ed McGrath, Cold Weather Project 2020, P.O. Box 81961, College, Alaska 99708).

His bottom line on this is the same thing. You should give extraordinary care, concern, and attention to detail in your vapor barrier, worrying very much about every penetration, every electrical outlet, every corner post and ceiling-to-wall junction. Because if you really put in a decent vapor barrier you're going to solve an awful lot of your insulation and other problems, and he sketches out ways of doing this.

One other thing, I talked to the people who build cold storage rooms. They don't worry about insulation at all; they just worry about vapor barriers. If they put in a decent vapor barrier, the rest takes care of itself. What they use is 6 mil aluminum foil with 3 inch wide foil tape. They are very careful to seal over every joint after it's been nailed so they cover not only the joint but the nail. So I'm beginning to believe that the vapor barrier is just as important as the insulation, particularly the integrity of the vapor barrier Ideally, we would like to have an unblemished vapor-impermeable skin inside the insulation.

I-6 Where do you put the vapor barrier in an envelope house?

--- The inner wall is in a highly moist atmosphere provided both from human beings and from plants in the greenhouse, but the dew point temperature is not going to be reached in the inner insulation, so I'm not worried about a vapor barrier to protect that inner insulation from a dew point condition. The vapor barrier is logically on the interior surface of the outer envelope, both walls and ceiling.

DC- The absolute humidity in the house and in the envelope will always be virtually the same, due to the constant air flow. The differences in relative humidity are based on temperature. Sometimes the envelope is warmer and sometimes the inner house is warmer, so at different times the vapor pressure will be going both ways. It seems best to let the inner wall breathe.

HH- We have come to the conclusion that the dew point won't be reached in the inner wall, therefore, leave out the vapor barrier in the inner wall.

--- Just make the vapor barrier very carefully continuous on the inner surface of the outer wall.

I-7 How do you make a vapor barrier "very continuous"?

BH- It's mainly a matter of diligence. Use the thicker 6 mil plastic, for one thing. Overlap it across two studs at the end of a piece, caulk the edges around windows and doors and where the barrier meets a sill plate or foundation wall. Keep it tight around electrical boxes, using tape. Better yet, design it into your framing system like Bill Brodhead has done so that you won't have those breaches to repair.

J. FIRE PROTECTION

A firefighter, writing in the Maryland Fire & Rescue Bulletin, saw it this way: "Any fire within the 'convection envelope' could be expected to spread rapidly to involve the entire dwelling. Persons within would find themselves inside an oven, with escape limited to the side walls if these contain usable openings..." As one conference participant put it, "You don't want to open the Boston Globe three days from now and find a front page lead article on some house that went up. A lot of the reason everybody's here is because you've had fantastic media coverage of these things. You'll also get fantastic media coverage if one of them burns up in 15 minutes."

J-1 What types of fire protection are being built into envelope houses? How are they designed to work?

JB- Like any house, it should have smoke and heat detectors to warn of danger. Like any house, ventilation control dampers should automatically close in case of fire. The superior communication characteristics of the loop will guarantee the quickest possible detection of excessive heat or smoke. Because of its loop nature, all the air in the loop must pass a given detector location, unlike a conventional house which depends on a fire to drive its product to a remote detector.

BH- A number of builders have site-built drop-down fire dampers that are held open by low-melting fusible links so that if the temperature in, say, the north plenum were to get abnormally hot, the dampers would drop down and block air flow all along the loop at that point.

PP- I think these dampers should be interconnected, so that if one door trips, they all close. If you have an 8' wide door with a fire under it, the path of the hot gases will immediately run around the edge of the door, never reaching the next link.

--- We try to keep the fusible links at such a low threshold that they trip pretty easily. I use 165° fuses.

--- 140° here.

--- The dampers will not stop a fire, so all you're asking them to do is give the owners time enough to get out. The smoke detectors are going to do a lot more towards insuring that than the dampers will.

--- If you use firecode sheetrock on both sides of the plenum, that might be more beneficial in the long run, with good smoke detectors, than trying to hassle around with these dampers.

--- Honeywell also makes a unit that interconnects all the detectors.

J-2 How are fire dampers constructed, and where should they be located?

BB- The fire dampers we've used have been of 5/8" drywall with metal channels around the perimeter to make them rigid. Drywall laid in by itself will warp and twist when subjected to high temperatures, so you need to keep it rigid. Then the fusible links are attached so that the links are exposed to the air.

WM- If you go to any good heating and air-conditioning shop, they can get you the links with no trouble; they're very inexpensive. The important thing is to make sure the links are in series, so that when one goes, they all go. We connected all the links along the north plenum with a wire that takes a Z-bend at each link; the links are attached to the diagonals of the Z's, so that if one damper falls, it pulls all the others down.

BB- If you are going to use fire dampers, they should be positioned above something flammable.

In other words, if you have a basement, there's really not much reason to put a fire damper at the bottom of the first floor, since there is not that much that's flammable underneath. Fire tends to move upward very quickly, so I would put a row of dampers at the bottom of the second floor, and if you want to do a super job, then also at the corner of the attic where the north wall reaches the attic rafters.

J-3 Any additional thoughts on fire protection or prevention?

--- In some of the earlier designs they used Styrofoam on the wall; that could be a real problem. And probably the worst thing you could do for fire would be to leave the polyethylene exposed--if you blinked, you would miss the flame shooting from the bottom to the top of the wall.

JF- I saw a fire during construction, and it was something. The plumber was under there brazing some pipes together, and the wind just caught the flame; it whipped about two feet up the north wall, and the kraft paper started to blaze. He grabbed it as fast as he could and started pulling the insulation out. He is a man of prayer, as well.

J-4 Has anyone had any problems with fire code officials over their designs?

DH- I talked to our new building commissioner. The last five years he spent as one of the technical staff, writing the new Massachusetts code. He spent 30 minutes looking at the plans and said I had no problem; we have fire-dampers in the north wall plenum at the first floor ceiling level.

DB- We showed the Sant Bani School envelope to the New Hampshire fire marshal, and he said that with firecode gypsum on both sides of the inner wall he considered us safe.

--- Our fire chief came back and said, "Hey, the more I think about this, you're probably better off than most of the contemporary houses around here that have 1½ or 2-story high living rooms or whatnot; they've got just the same vertical draft space, and they don't have a fire detector at the top of every high space."

K. AIR QUALITY

K-1 Does an envelope house have an earthy odor?

GB- It's a wonderful smell. I just open up the greenhouse doors and--it's great.

BB- I actually have a client who's allergic to soil. You can put a skim coat of concrete over a crawl space without losing any of the effectiveness of the mass. as far as heat transfer goes.

K-2 How do you deal with humidity, with all those plants?

JB- The earth works well as a humidifying store and control, primarily because its moisture content is much greater than that in the house air. Thus it can absorb excessive moisture, or give up desired moisture, without excessive dampness or standing water. The result is a self-controlled humidity which protects and extends the life of the building materials, not to mention the human inhabitants.

DB- We're planning a house for some people with asthma who see the humidity control feature as one of the big plusses of the envelope house.

HH- Access to ground moisture in combination with active plant growth helps maintain relative humidities at adequate levels year round. In our house we have typically 40-50% relative humidity through the winter.

BH- A number of houses have gotten excessively wet just before being enclosed; that, plus the moisture that comes out of new concrete have made some of the houses more humid initially than we believe they will be when their natural balance is established.

K-3 What about internal pollution?

WM- A very serious problem in these super-tight houses is the internal pollution caused by fiberglass particles in the air or fumes from urea formaldehyde or carbon monoxide from a gas stove. One of Lee Butler's original feelings about the envelope house was that the plants in the greenhouse would improve the air quality by absorbing some of the carbon dioxide and carbon monoxide and giving back oxygen. Any evidence of that, Hank?

HH- Well, it's always refreshing to step into the greenhouse. One thing the greenhouse won't remove is radon gas, which has been found to be coming out of the earth and particularly out of masonry products, brick and concrete. It turns out that the levels of radon and some of these other high-tech pollutants are alarmingly high in very tight houses and especially in super-insulated houses. There are ways of getting around the problem, and you can get as fancy or as simple as you want with air-to-air heat exchangers, which take the heat from stale air going out and warm the fresh air coming in. Heat exchangers are also a way to control humidity levels.

BH- Just plain stale air from human beings and especially smoking beings is an inescapable result of making a house as tight as we want to make it. Heat exchangers will definitely do the trick, but I just want to point out that at any time the envelope can be treated as a large air-lock, which can be completely aired out without significantly chilling the inner environment. The heat capacity of that air is quite small--the heat's mostly in the earth and the drywall and so forth--so not that much is lost if you vent it for ten minutes on a sunny winter day. Then, after the envelope's come up to temperature again, which won't take long, open the inner house to the envelope for a spring cleaning of the air.

K-4 What does anyone know about heat exchangers?

DH- There's a report from Lawrence Berkeley Labs which did a whole study on the Mitsubishi model and found that they saved considerable amounts of energy--40 million BTU's a year in certain circumstances.

RS- A point about air-to-air heat exchangers. The Mitsubishi is very efficient at transferring latent heat as well as sensible heat. Now what you're looking for in a lot of these installations is a device that transfers sensible heat without transferring latent heat. Which means, for instance, that instead of a paper heat transfer surface, you'd have a plastic heat transfer surface.

--- The Mitsubishi unit, which is number VL-1500-MC, works great in the summertime because of the fact that it does exchange latent heat. By dumping the heat latent in the air's moisture outside, it helps to keep the house drier as well as cooler. But in the wintertime you're dumping all the moisture back into the house.

BH- That's desirable in a normal house, which gets very dry in the winter, but the envelope probably doesn't need it--which isn't to say that it couldn't handle it.

--- Has anyone around here built this Canadian polyethylene film heat exchanger? The one out of Saskatchewan.

--- LBL looked at that one in the same study. They commented that it didn't exchange latent heat, and they were most interested in the cooling aspects.

(A copy of the bulletin describing its construction can be had by writing for "An Air-to-Air Heat Exchanger for Residences" from the Department of Mineral Resources, 1404 Toronto Dominion Building, Regina, Saskatchewan S4P 3P5).

K-5 What about venting bathrooms and household appliances, such as dryers and gas stoves?

DD- In both the houses I've done I vented the appliances, dryers for example, into the envelope; the bathrooms were vented directly to the outside.

JF - We did it just the opposite; we vented the dryers to the outside and the bathrooms to the envelope. There is no noticeable odor or anything from the bathrooms.

DC- We vented the dryer, kitchen, and bathrooms all into the envelope.

L. SPECIAL CONCERNS

L-1 Is there anything that can be done to prevent the condensation on the north and south windows?

JF- I don't think so. One thing you can do to make it less of a problem is to take an aluminum J-channel and adhere it to the bottom of the glass with silicone sealant so that it catches the condensation as it comes down off the glass. It will evaporate during the course of the day; this will prevent it from discoloring or rotting the sill.

DC- I haven't seen this, but it seems to me you could slope the sill to drain the water away in situations where you anticipate moisture problems.

BB- Some of the envelopes have had very little condensation and others have had considerable condensation. The difference seems to be the amount of water in the soil itself. If the house is in a very high water table, careful consideration has to be given to removing that water, possibly even with a vapor barrier underneath the crawl space earth. If the ground is relatively dry, it will actually be removing moisture from the air. Concerning the condensation, we have considered using triple glazing and storm sash on the north side to help reduce the temperature difference so that we get less condensation.

L-2 Is anyone concerned that the condensation may rot the sheetrock or the wood?

HH- Actually, the only place that really has condensation in the envelope is on the glazing. I haven't noticed any other condensation, even though when we first moved into our house our average relative humidity was running about 70% the first winter because of all the moisture trapped in the concrete block and the earth packed down in the greenhouse. There was no moisture condensing on the insulation or the studding or anywhere except on the windows. So treat that as an area of high moisture and be done with it.

DB- We used aluminum thermal-break windows with synthetic marble sills.

L-3 What about noise? What are the acoustical characteristics of an envelope house?

GB- I don't have any windows on the north side of my house and I find it very frustrating because the house is so quiet I can't hear what's going on outside.

BH- If you were designing a house primarily from the standpoint of isolating yourself from external noise, this is the house you'd get.

L-4 With all that earth so close to the framing, is any special termite protection in order?

BH- The closeness to the earth is no different from any other house with a crawl space, except that the crawl space is ventilated through the envelope, rather than directly to the outside. Still, the usual precautions that are taken in you area of the country should be needed.

JF- The owner of the house I built was very interested in intensive gardening, and for that reason he didn't put any termite protection down other than termite shields, because he was afraid of the leaching of the chemicals into his greenhouse.

L-5 How does one keep rodents, pregnant cats, etc. out of the envelope?

BB- You can screen off your cooling tube with a metal screen or hardware cloth. If you're worried about animals getting up in the envelope, ½" screening should be adequate to keep a mouse out.

JB- You have a lot better chance to control or eliminate pests in a loop region than you do in your backyard or garage. If you like bugs in your greenhouse, but not in your inner house, then add some screens. Why not birds in your solarium? In the future will we have bird doors, as we now have cat and dog doors? One person's pest is another person's pet.

M. COSTS

M-1 How much have these envelope houses cost, and how does that compare with comparable non-envelope housing?

DD- For the two I've done--the cost on the first one was $33 per square foot of finished space, and the second one was $35 per square foot of finished space. Both included improvements on the land, but not the price of the land itself. The envelope houses cost about 10% more than the bermed houses I was building.

DB- When I look at the elements, yes, the double walls cost more, but I think for a total house it's not that much different--$35-40 per square foot. However, when people come to me who can't afford more than $25-30,000 for a house that is that big, I don't do an envelope.

HH- The cost per square foot of building an envelope house in Peterboro is comparable to the cost per square foot of conventional.

--- We're getting $35 per square foot for relatively standard stud-framed homes.

BB- I think the first thing you have to realize is that you can't compare the envelopes you have seen with tract homes that come out of a box and go up in two weeks. What you are actually seeing are custom homes, and if you compare then with custom homes, you'll find the cost per square foot to be very similar. Where you have a higher cost factor, though, is the space in the envelope house that is devoted to the north wall space and to the greenhouse. So either the owners have to try to utilize those spaces to a much higher degree or increase the size of the house to get the same interior space.

<u>M-2 What thoughts have the designers had toward holding down the cost of this type of housing?</u>

DD- You have to reconceptualize your building and you have to reorganize the structure and layout of the building to accept the envelope. If you take a stock plan of a saltbox or something and try to add on an envelope, you'll be wasting a lot of money. The way to trim the costs is to think it through initially and come up with an envelope design and not an adaptation.

BH- I think there are a number of interesting ways to attack this problem. You can figure out creative ways to make usable space out of the envelope, even though it's not space that's 65° for 24 hours a day: the northside airlock entry, stairways and storage space--a workshop space would be excellent. The other approach I've seen is in Don's gambrel design, cutting the envelope volume down dramatically to where it just covers the windows. If that turns out to be the major contribution the envelope makes, reducing losses through the glazing, you'll be seeing more hybrid designs, part envelope and part super-insulated.

DC- I think once people get a better understanding of what they're really trying to accomplish, they'll see ways to streamline the design, to cut out some of the unnecessary stuff. Why have two frames, if only one is holding up the house? I'm thinking of something like those stressed skin panels for light, even moveable partitions, and suspended insulating ceilings. Another thought is that forced convection may be so effective that you'll want to leave out the north envelope and just super-insulate there.

<u>M-3 Where do you see the envelope concept going in the future?</u>

GT- We now have a situation where less than 5% of the population can afford single-family housing. Here we are in New Hampshire, a unique place, where a lot of people want to get out into the hills and burn wood and that sort of thing. You've got to put yourself into a larger perspective here and realize that you're talking, so far, about luxury housing. How is the envelope concept going to be adapted to what's ultimately going to have to happen to the housing industry, namely condominiums, cluster housing, and factory-made housing?

MK- I think you make a big mistake if you think that this greenhouse is a thermal artifact. Customers come to us--nobody asks any more for passive houses or even solar houses; they want an envelope house, and the reason, I think, is that we've finally gotten together what some architects were trying to do in 1939, which is design a house that has a lot of glass on it, but instead of it being an enormous heat loss and a privacy problem, you now have something that uses almost no heat and gives you a barrier, a sort of semi-private intermediate space. If we can get that together with the need for lower cost housing, I think we've really got something going.

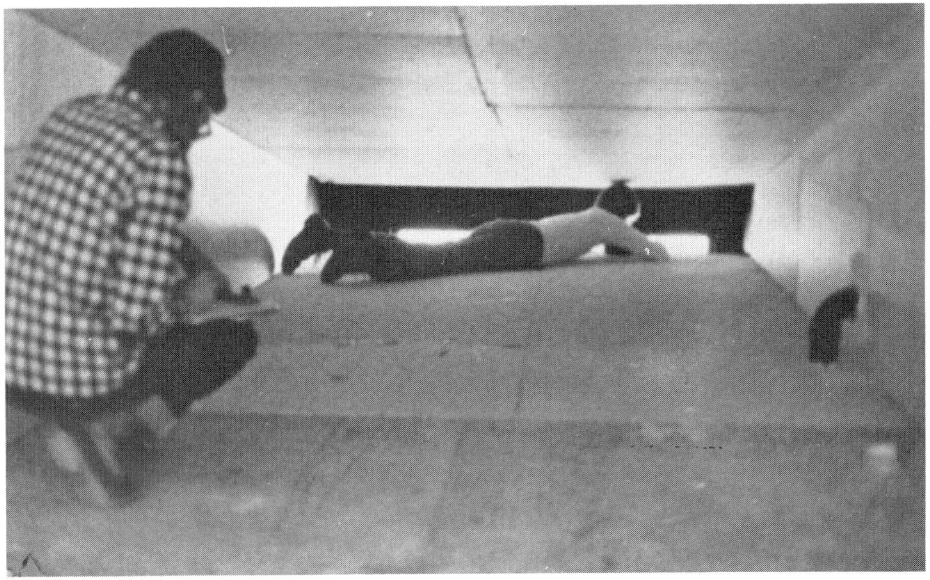

Doug Clayton and Hank Huber checking airflow in Burns attic with pitot tube

Charles Seaver, Robert Smith, Norman Saunders, Doug Clayton discuss data collected at Burns

Stereatronics data logger as installed in the Burns basement

7. PERFORMANCE by Doug Clayton

In this chapter we will present several graphs of temperatures recorded at the Burns and Beale envelopes of Community Builders and the three envelopes of Hank Huber. Most of the data was recorded at the Burns house with a Stereatronics Model 78101 data logger built by Norman Saunders and purchased with funds from a D.O.E. Appropriate Technology Grant. The Logger also recorded for brief periods at the Beale and Huber's Sharon houses.

Al Converse and Art Blake of Dartmouth College also monitored envelope temperatures, air flow and solation at the Burns with a Leeds and Northrup strip chart recorder from 2/8 to 2/16/80. One "typical" day's data is presented from this record.

Also presented are some graphs and a chart of temperatures prepared by Hank Huber from manually recorded temperatures at his three envelope houses. Of particular interest here is the performance of envelopes with forced convection.

The Stereatronics data logger recorded temperatures and solation at the Burns during the period 10/17/79 to 2/4/80. The logger is essentially a multiplexer that scans 30 odd analog channels. Also on board are 8 digital channels, two of which are used to keep time and one to integrate solation. The temperature sensors are solid state and consist of diodes and filters. They are excited by a small current and return a linear voltage drop with increasing temperature. The solation sensor is a photo cell mounted vertically inside the south glass. The data, through a timer, is put on a cassette tape once an hour in digital form. A week's worth of data fits on 45 minutes of cassette tape. The tapes are sent to Norman Saunders for processing on a small computer. We then manually graph the computer print-outs. Saunders can also display the data on a cathode ray tube.

OCTOBER 17 TO JANUARY 28

The first graph is a max-min temperature bar graph covering 3½ months at the Burns house. The sensors graphed are the living room, greenhouse and outside (ambient) air temperatures, surface and six inch deep earth temperatures and solation. Blanks in the data are due to equipment or operator failure.

The living room temperature sensor was located in the southwest corner of that room, out of the direct sunshine at shoulder level. It can be seen that the house interior frequently

7.3

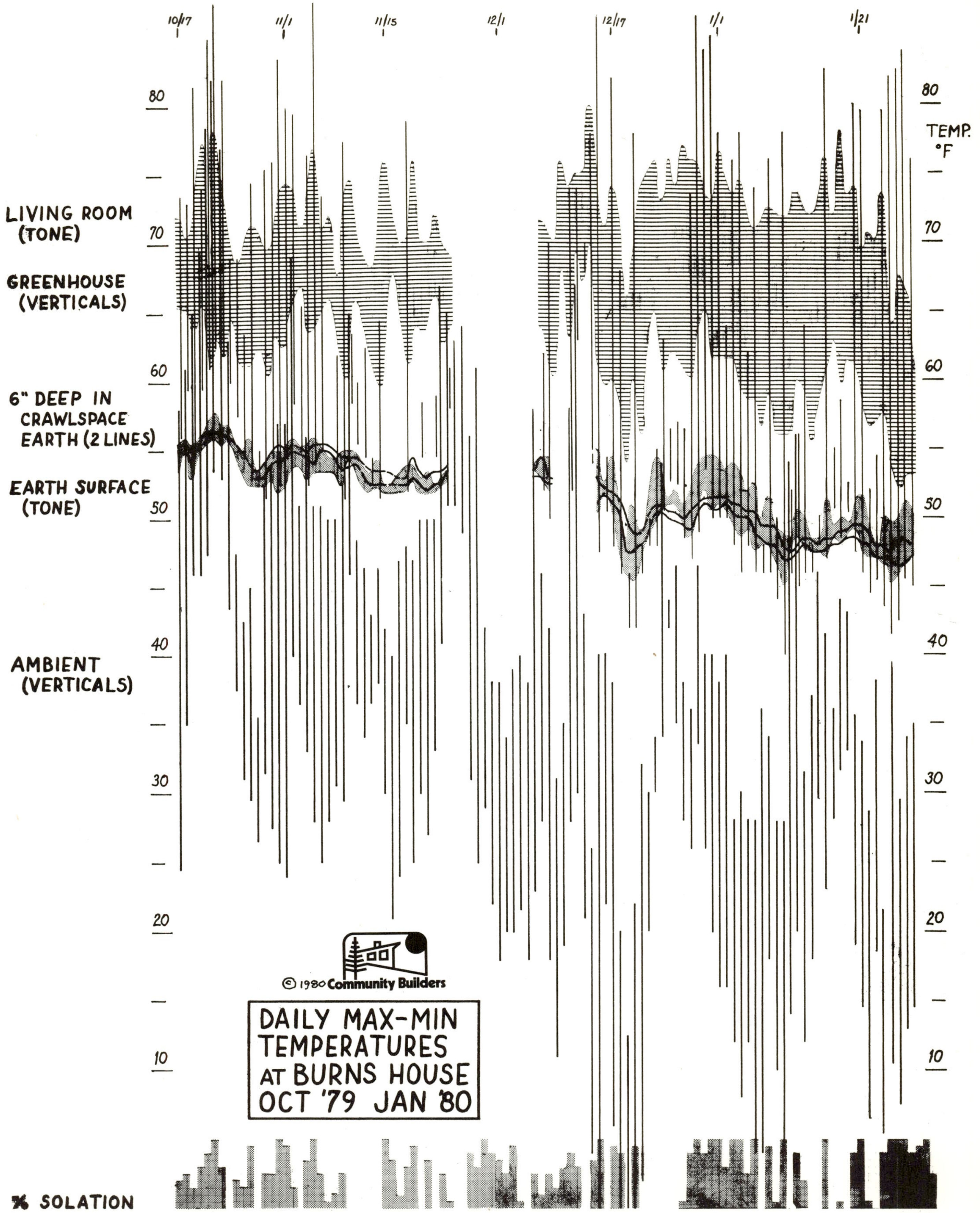

10/17
11/1
11/15
12/1
12/17
1/1
1/21
80
70
60
50
40
30
20
10
TEMP. °F
LIVING ROOM (TONE)
GREENHOUSE (VERTICALS)
6" DEEP IN CRAWLSPACE EARTH (2 LINES)
EARTH SURFACE (TONE)
AMBIENT (VERTICALS)
% SOLATION
© 1980 Community Builders
DAILY MAX-MIN
TEMPERATURES
AT BURNS HOUSE
OCT '79 JAN '80

drops to 60°F and occasionally below in times of stress (little sun and cold outside). The coldest interior temperatures occurred at the end of January when the house was unoccupied and didn't benefit from the normal internal gains or convective gains from the greenhouse. You will notice in the more detailed graphs that follow that the inner house temperature was generally above 65°F between noon and midnight.

The greenhouse sensor graphed here was located about seven feet above the floor near the south wall and out of the direct sun. On sunny, not too cold days, the greenhouse would climb above 80°F. On diffuse days temperatures hovered around the mid 50's. The greenhouse typically reached lows in the mid 40's by January.

The earth surface and 6 inch deep sensors graphed were located in the west central portion of the crawl space 12 feet from the north wall. Of course the average earth temperatures decline through the winter, but notice the rapid recoveries after periods of stress. The earth surface is typically experiencing a 4-6 degree F diurnal swing while the 6 inch depth is swinging 1-2°F. These earth temperatures were verified to be representative by other sensors and by manual temperature readings. Max-min mass temperatures throughout the crawlspace were remarkably uniform. The basement floor surface experiences about half the amplitude of the crawl space surface. We have speculated that this is due to the foil face on the ceiling above. (See page 7.8).

Each inch of depth of the crawl space earth and north concrete wall contains about 110 ft^3 of mass. With an assumed average heat capacity of 35 $BTU/ft^3 \cdot {}^oF$ (moist earth) each inch would have a heat capacity of: 110x35=3850 $BTU/ft^3 \cdot {}^oF$. The energy stored or released by a 5°F temperature change at the surface and 1°F change at the 6" depth (with a non-linear distribution between) would be: (3850 BTU/°F x 5°) + (3850 x 3.1) + (3850 x 1.9) + (3850 x 1.3) + (3850 x 1.1) + (3850 x 1) = 51,590 BTU. There is also some latent heat transfer occuring through the daily cycle of condensation and frost appearing on the outside glass at night that thaws and evaporates in the morning.

We regret that we are unable to provide data from our deeper earth sensors. We experienced calibration problems and equipment and technique failures and are not confident in the data we have. Norman Saunders commented at the Feb. 23rd conference that "the ground temperature dropped from 65 to 50 degrees F in the course of three months and that presumably went down no deeper than ten feet." Our deepest sensors were about 5 feet below the crawl space surface. 5 feet of depth under the Burns house contains about 6,000 ft^3 of earth. With a heat capacity of 35 $BTU/ft^3 \cdot {}^oF$ and a 15°F temperature change, the discharge would be: 6,000 x 35 x 15 = 3,150,000 BTU.

DECEMBER 17 THROUGH DECEMBER 21

Three more temperature sensors are represented here in addition to the five in the previous graph. The peak sensor is located in the middle of the 11 inch air space where the greenhouse connects with the roof envelope. The north envelope sensor is in the west side of the north plenum centered in the eight inch air space two feet above floor level. The

7.5

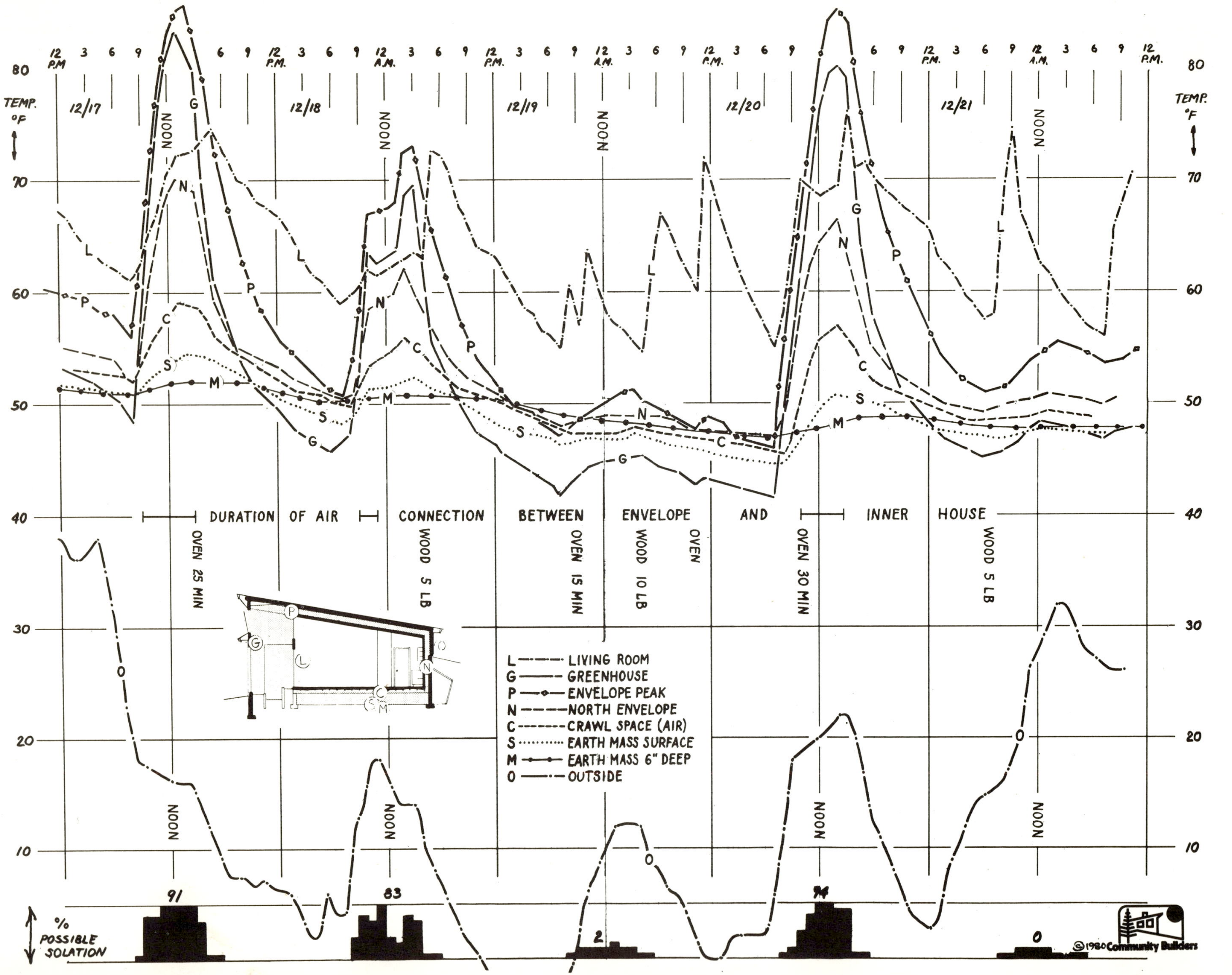
TEMP. °F
12/17
12/18
12/19
12/20
12/21
NOON
DURATION OF AIR CONNECTION BETWEEN ENVELOPE AND INNER HOUSE
OVEN 25 MIN
WOOD 5 LB
OVEN 15 MIN
WOOD 10 LB
OVEN
OVEN 30 MIN
WOOD 5 LB
L LIVING ROOM
G GREENHOUSE
P ENVELOPE PEAK
N NORTH ENVELOPE
C CRAWL SPACE (AIR)
S EARTH MASS SURFACE
M EARTH MASS 6" DEEP
O OUTSIDE
% POSSIBLE SOLATION
91
83
2
94
0
©1980 Community Builders

crawl space air sensor is located one inch below the Masonite material which is covering the underside of the floor framing, directly above the earth sensors.

Of the many things to be seen in this graph, probably what stands out most is the ineffectiveness of the envelope to move a large percentage of the solar gain to the earth beneath the building. Notice that on sunny days the peak air temperature is 25 to 30 degrees F above that in the crawl space.

The solar gain on typical sunny winter days between 8 am and 4:00 pm would be:

1600 x 340 x 0.8 x 0.8 x 0.9 = 313,000 BTU/day

where:

1600 BTU/ft^2·day	radiation incident on one square foot of vertical surface
340 ft^2	aperture of the Burns house
0.8	transmittance of aperture
0.8	absorptance of interior
0.9	percent solar

The heat loss back out the glass and roof for the same 8 hour period would be:

350 x 60 x 8 = 168,000 BTU

where:

350 BTU/hr·$^{\circ}$F	conductance of the envelope to the outside
60°F	is the average temperature difference between the envelope and outside during the 8 hours of gain.
8	hours of gain.

Heat stored in the mass below the building on a typical day was around 50,000 BTU or 50,000/300,000 = 16%. Of the solar energy gathered by the envelope or 50,000/(300,000-168,000)= 37% of the total energy stored somewhere in the building. All these numbers are very significantly depending on the particular day and conditions being observed.

At night all envelope temperatures tend to converge around the 50°F earth temperature, except for the greenhouse whose loss through the south glass is driving the convective loop, pulling heat from the structure and earth.

Notice the difference in slope of the living room temperature following days of sunshine and days of cloudiness. The temperature declines approximately one degree F/hr after a sunny day and two degrees F/hr after a cloudy day. This seems obviously due to differences in energy storage in the structure of the building and a resultant greater Delta T between the inner house and the envelope after cloudy days.

The interior responds rapidly to small BTU inputs from cooking and the wood stove. This is a function of the small size, tightness and low mass of the space. Typically only two to ten pounds of wood were burned following sunless days and a total consumption for the heating season to date (4/18/80) has been less than 400 pounds or a tenth of a cord. Though the total auxiliary load is low, the occasional small spurts of energy required means that many fires must be built if wood is relied upon.

NORTH ENVELOPE AND CRAWL SPACE CROSS SECTIONS

These graphs show in greater detail the temperature distributions at these locations in.

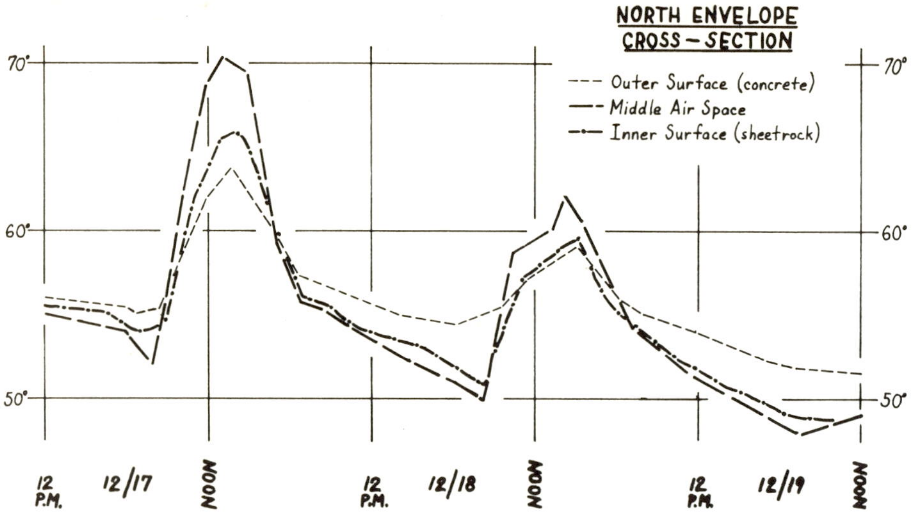

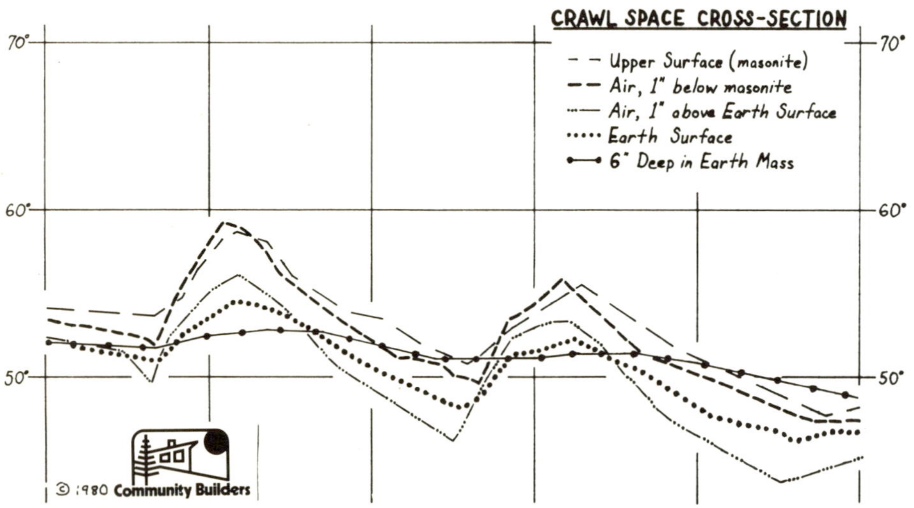

the envelope loop. In the north plenum notice that the concrete foundation wall experiences a more moderate temperature swing due to its massiveness and ability to take and give up heat. At night this concrete surface is radiating energy to the cooler sheetrocked outer surface of the inner wall. In other words, the envelope seems to be making very effective use of its earth berm.

The crawl space cross section clearly illustrates stratification of the air in the space as the air one inch below the masonite is two to three degrees warmer than that one inch above the earth surface. Though the sun-warmed air is hugging the bottom of the floor framing that apparently isn't a big problem because about half the heat transfer to the earth is by radiation (about 1 $BTU/ft^2 \cdot hr \cdot {}^oF$) across the gap.

EAST AND WEST LOOPS

This is probably one of the most interesting graphs we have done, though difficult to interpret. We think it is illustrative of some of Jim Berk's Rules of Thumb.

The west 24 feet of the Burns house has a cathedral ceiling and additional acrylic glazing above the single row of 92" thermopane windows. The envelope at the top here is an 11-inch air space formed by a double row of purlins between beams which are on eight foot centers. There is no north window above the west, north envelope sensor.

In contrast the east 16 feet of the house lacks the additional aperture above the glass units, the top of the envelope is formed by an attic with wider, more "rounded" connections between the vertical portions of the envelope and the attic portion. There is a north window above this north envelope sensor.

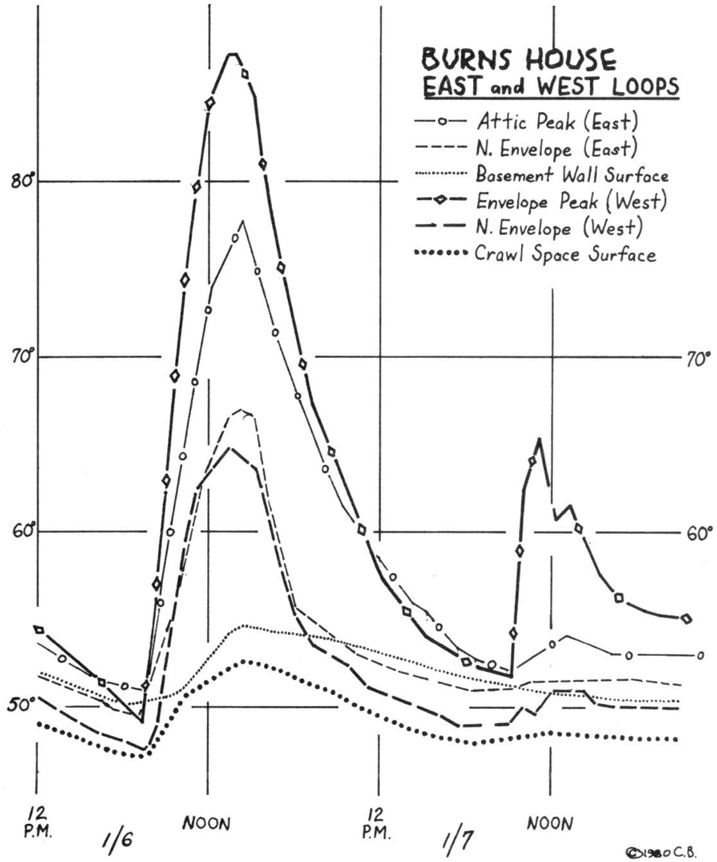

Notice that even though the envelope peak reaches a higher temperature at the west end of the building, the north envelope, even though below a window, gets warmer on the east side. The mass surface is also warmer in the east than in the west.

When air velocity measurements were taken at noon on a sunny day (December 7th) the area where the highest velocities were measured was below the north window on the east end of the building. Here the velocity was 40-45 ft/min, two to three times that measured anywhere else in the north plenum at that time, even below the other north window on the west end. (Total volume of air moving in the envelope was about 400 cfm, calculated from velocities at the most restricted point in the loop, the bottom of the north plenum).

Berk suggests that:

- --collection should be concentrated down low on the south side of the building
- --overglazing may result in loop strangulation and dumping of excess collection
- --there should be a grading of heat/loss on the north so that you provide more higher up (e.g.: north windows)
- --the greatest restrictions to flow are at the corners where pressure and momentum losses occur.

Clearly the east end of the Burns house conforms better to these guidelines and seems to be working a little better.

LOOP CUT OFF

The first two days graphed here show the house as it typically performs on sunny and overcast days. Although the data logger didn't catch it, there was obviously a short period of sun shine on the morning of January 7. On the third day, January 8th, the envelope flow was blocked at 1:00 pm by stapling 6 mil poly across the peak. Notice the immediate drop in all the temperatures around the loop, except in the greenhouse. The following day, the 9th, was overcast and temperature curves here should be contrasted to other overcast days (12/19, 12/21, and 1/7). When the plastic was removed there was an immediate jump in temperatures as a blast of warmed air moved around the loop. The greenhouse reached its low temperature to date on the morning of the tenth. Clearly the envelope does transport and store some heat energy!

7.9

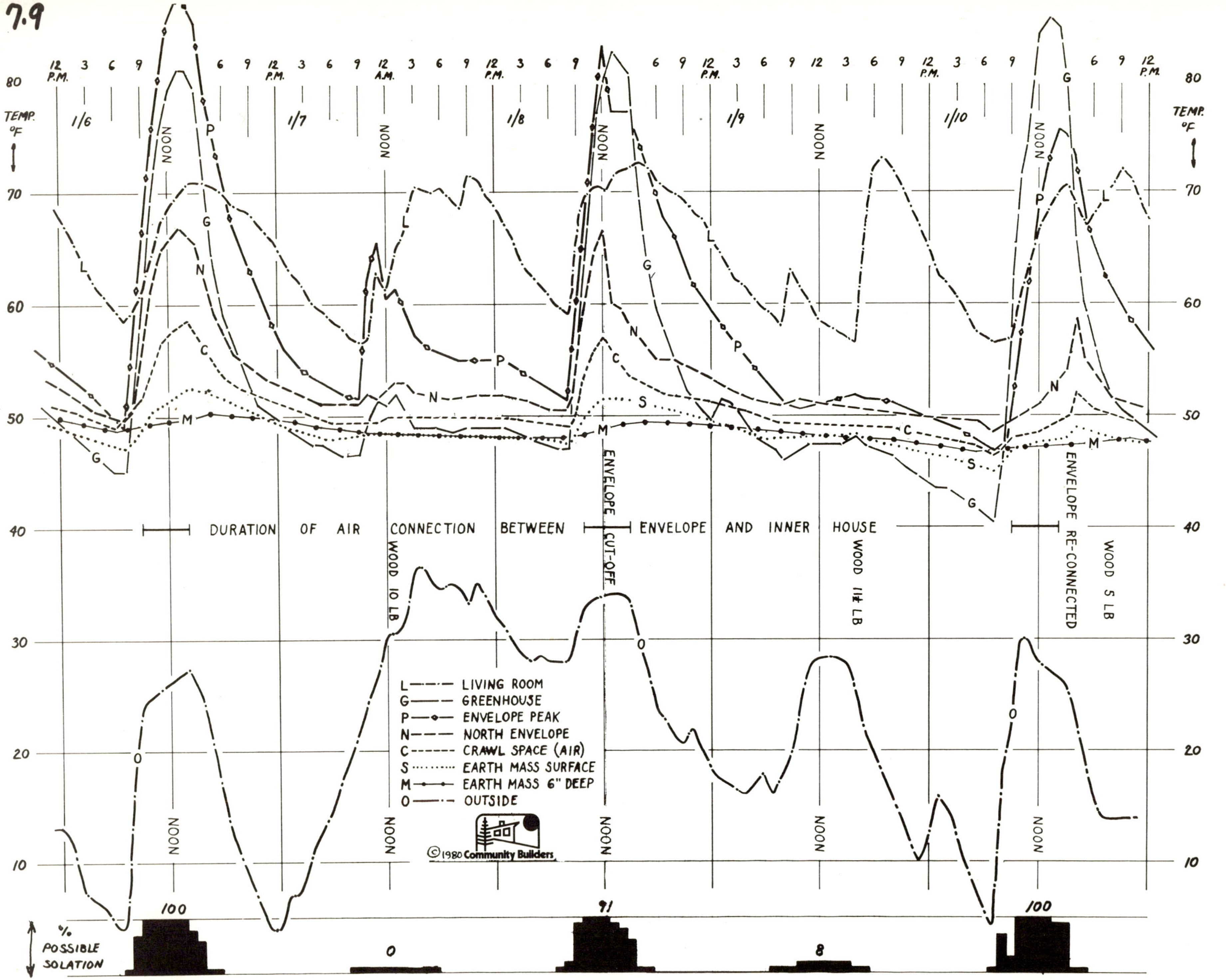

TEMP. °F
1/6
1/7
1/8
1/9
1/10
NOON
12 P.M.
12 A.M.
DURATION OF AIR CONNECTION BETWEEN ENVELOPE AND INNER HOUSE
ENVELOPE CUT-OFF
ENVELOPE RE-CONNECTED
WOOD 10 LB
WOOD 11½ LB
WOOD 5 LB
L LIVING ROOM
G GREENHOUSE
P ENVELOPE PEAK
N NORTH ENVELOPE
C CRAWL SPACE (AIR)
S EARTH MASS SURFACE
M EARTH MASS 6" DEEP
O OUTSIDE
© 1980 Community Builders
% POSSIBLE SOLATION
100
0
91
8
100

These comments on Al Converse's monitoring of the Burns house are paraphrased from the February 23rd Conference transcript.

AC- All the measurements we made were around the "east side" loop; they're all sort of in a plane. The insolation meter was mounted vertically inside the south glass. The hot wire anenometer (for air flow) was mounted below the center of the floor grate on the south. The rest of the measurements were temperatures around the east side loop, taken with thermocouples.

We were very interested in the fact that all the temperatures converged around 50°F during the nighttime hours. We also were interested to note that the air flow picked up almost immediately after the sun came up, ran fairly constantly through the day, and fell along with the sun as it set. There then was a period during the evening in which there appeared to be no air flow, at least at the point where we were measuring it, and then later in the evening, particularly after 12:00 midnight, there was noisy fluctuating air flow. The calibration of the hot wire is non-linear. At night the variation in velocity is from 10 to 20 fpm, in the daytime it is from 34 to 40 fpm.

DH- May I comment about that a little bit? I spent one night in that house with several of the other people in this room and a good part of that time I spent exactly under the grate that Al referred to, with incense sticks, and tried to map the flow across the grate... standing up over your head it's maybe 18" or something like that, from the bedroom wall to the outside south wall. At 5 or 6 in the morning there's a very decided downflow. There's cold air spilling out of the greenhouse through that grate into the basement. About in the middle where Al's sensor is there is almost no flow, and if you come back right against the last few holes up against the basement ceiling there is an upward flow, the hottest air in the basement rising to the basement ceiling is seeking to get out and it's creeping around that corner and going up. But that was only in the last two holes in that grate, maybe 3/4 of an inch. So there's a very complex flow situation taking place there in the middle of the night which is doing what I think all of our physics says should happen. The hottest air, which is at the ceiling of the basement is rising and the coldest air in the greenhouse along the glass is descending.

DC- I would like to point out the flatness of that basement floor temperature curve. I think we would be seeing a better thermal coupling there if the ceiling of the basement didn't have that foil-faced insulation on it and had a better emittance to the floor. I think it is also interesting how quickly the air velocity reaches its maximum, long before the insolation peak.

7.11

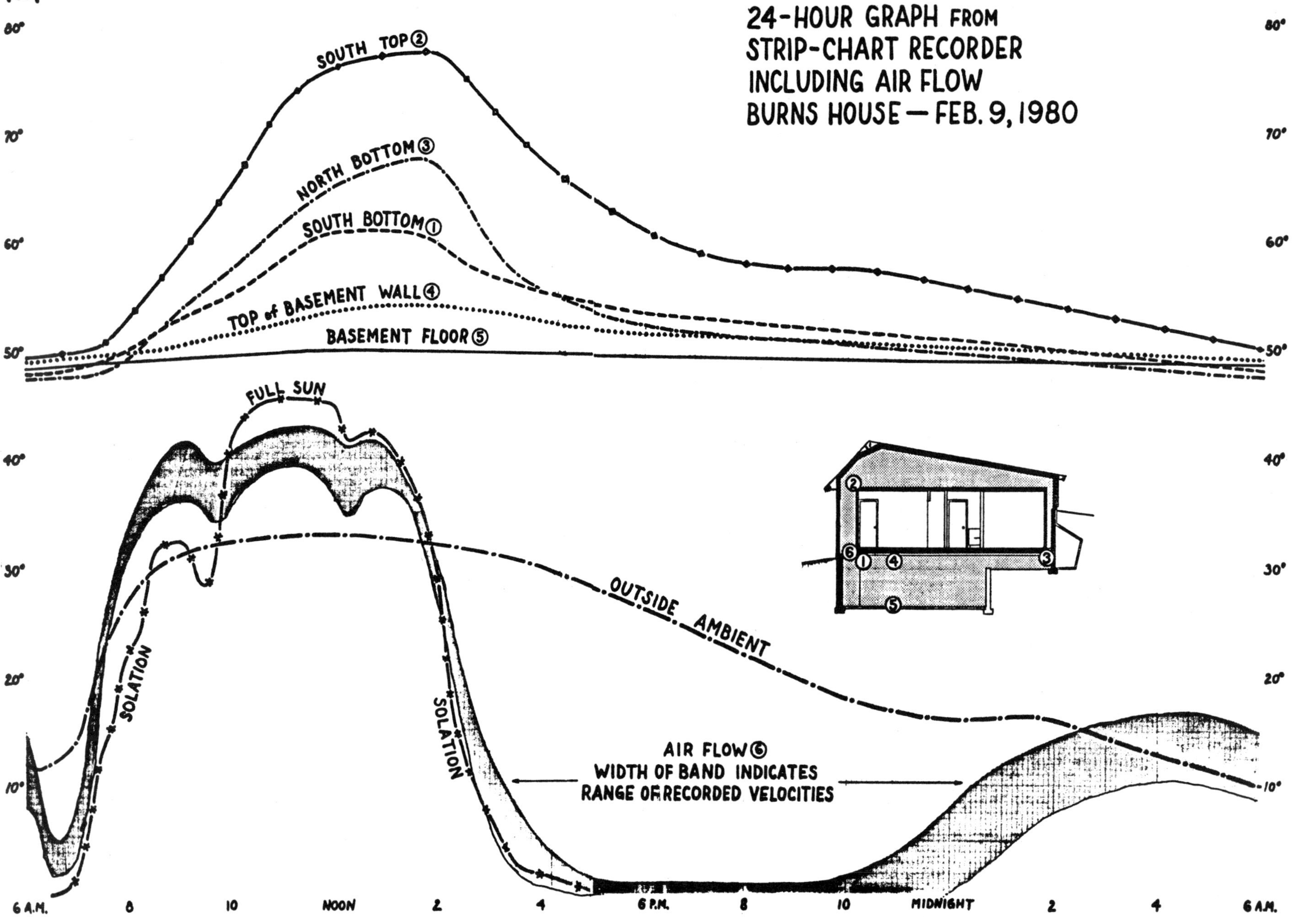
24-HOUR GRAPH FROM
STRIP-CHART RECORDER
INCLUDING AIR FLOW
BURNS HOUSE — FEB. 9, 1980
SOUTH TOP ②
NORTH BOTTOM ③
SOUTH BOTTOM ①
TOP of BASEMENT WALL ④
BASEMENT FLOOR ⑤
FULL SUN
SOLATION
SOLATION
OUTSIDE AMBIENT
AIR FLOW ⑥
WIDTH OF BAND INDICATES
RANGE OF RECORDED VELOCITIES
80°
70°
60°
50°
40°
30°
20°
10°
6 A.M.
8
10
NOON
2
4
6 P.M.
8
10
MIDNIGHT
2
4
6 A.M.

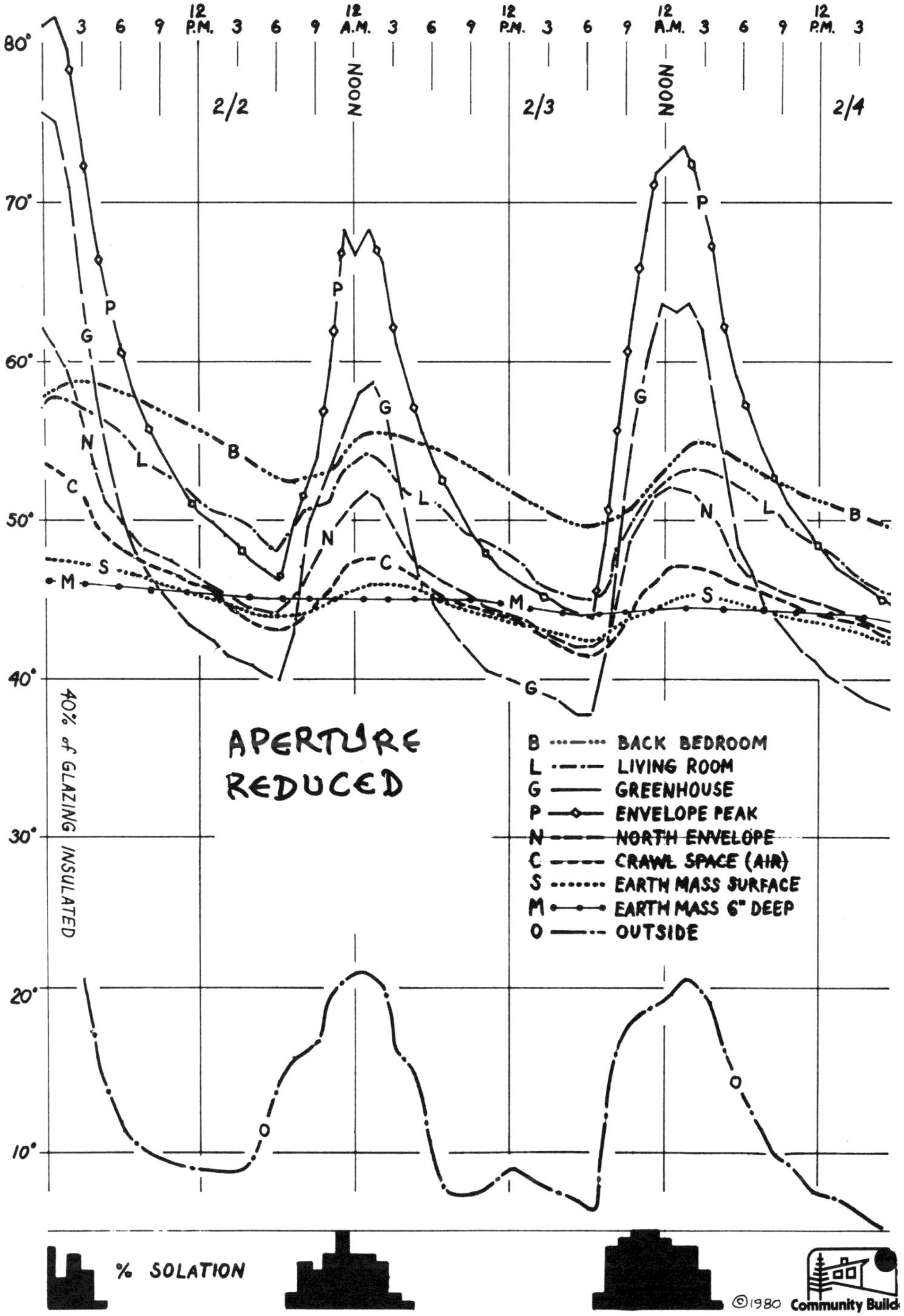

APERTURE REDUCTION

What we were trying to learn with this experiment was how well the house would perform with a smaller amount of south glazing. We accomplished a 40% reduction in the glazing by tightly affixing two inches of Styrofoam over half the outside of the glass units with battens and masking tape. Heat loss through the Styrofoam-covered glass was about R-13, through insulated wall areas about R-22. The Burns house has a ratio of glass to interior floor area of 0.29. This was reduced to 0.17. Unfortunately due to several factors the results are a little hard to interpret. The house had been unoccupied for nine days and hadn't experienced normal internal or convective gains, the data logger had been recording at another house so we had only a brief record of this period of unoccupancy, and we discontinued the experiment because the Burns returned from their vacation. Still, as we look at the temperature plots it does appear that the mean temperatures were lower with the 40% reduction.

We moved the data logger for brief periods to the Beale and Huber-Sharon houses. The graphs from those houses are presented also. We do not have wood consumption and oven use records from these periods. Still, it is apparent when large internal gains occurred from the jumps in recorded internal temperatures.

7.13

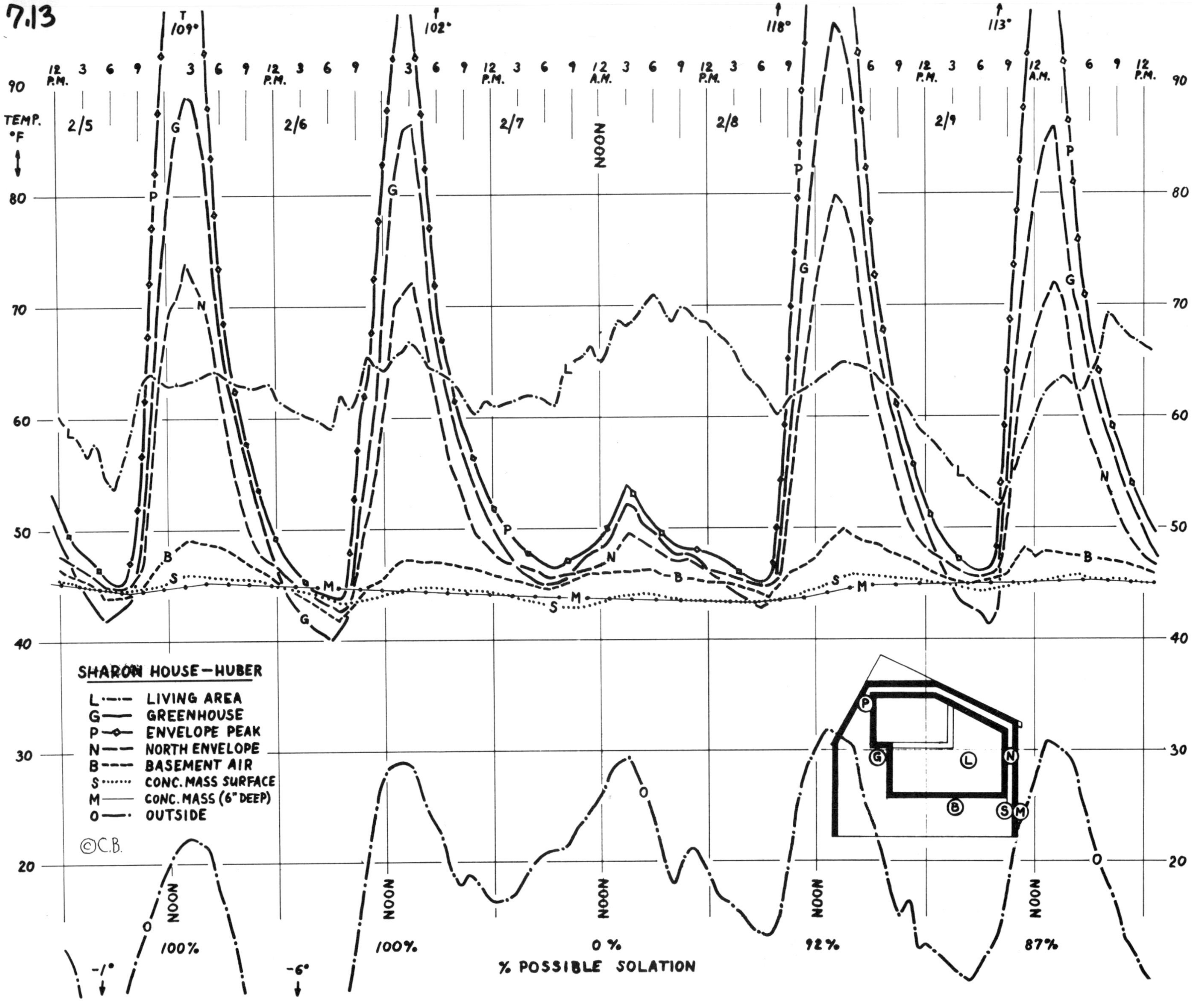

SHARON HOUSE—HUBER
L LIVING AREA
G GREENHOUSE
P ENVELOPE PEAK
N NORTH ENVELOPE
B BASEMENT AIR
S CONC. MASS SURFACE
M CONC. MASS (6" DEEP)
O OUTSIDE
©C.B.
TEMP. °F
2/5
2/6
2/7
2/8
2/9
109°
102°
118°
113°
NOON
100%
100%
0%
92%
87%
% POSSIBLE SOLATION
-1°
-6°

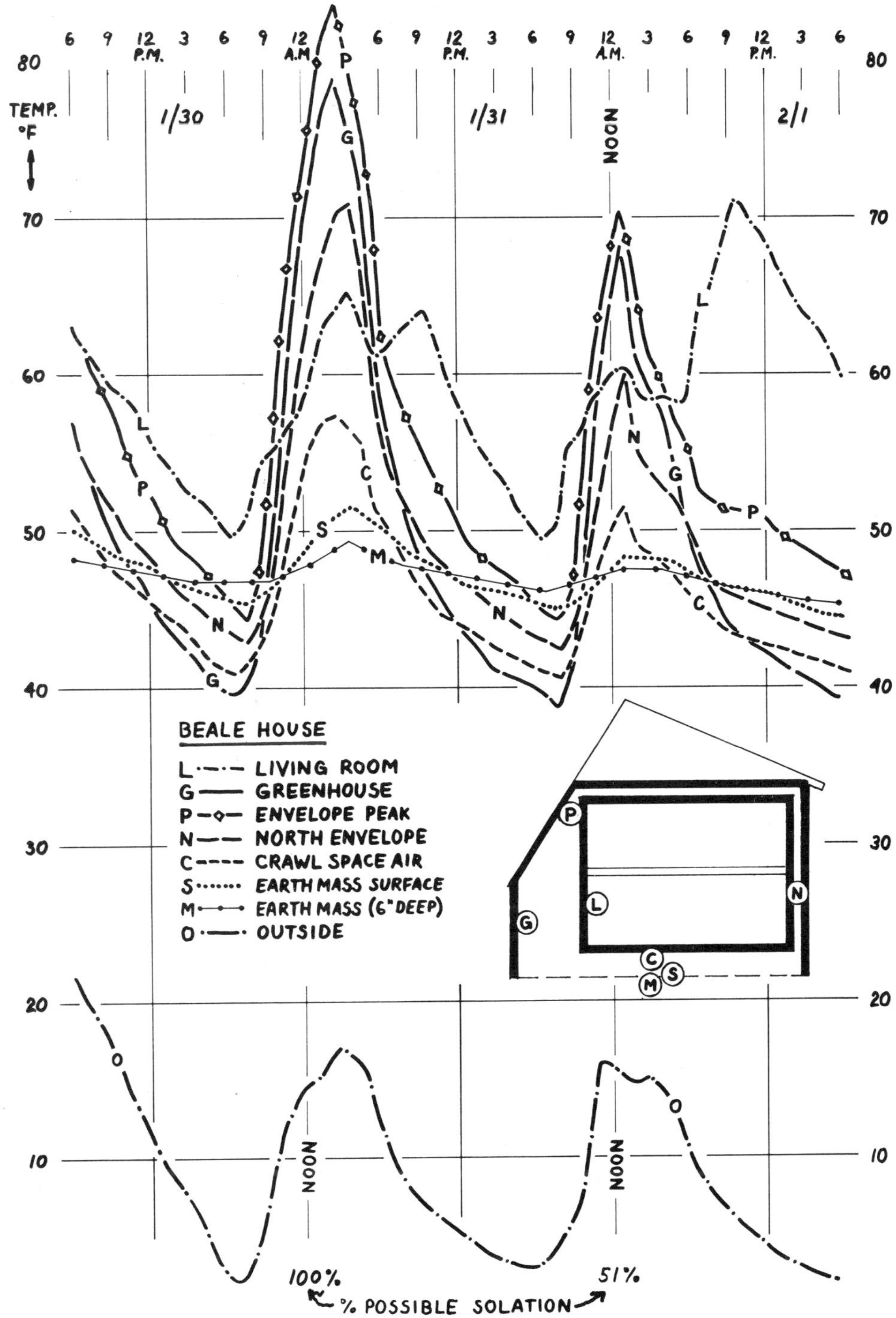

One apparent contrast among the houses can be noted. The temperature curves of the Burns and Beale house are quite similar to each other while the Huber-Sharon house has much greater amplitudes "above waste level" and less amplitude below (basement and mass) than have the Burns and Beale houses. We did not record at the Huber-Peterboro house with forced convection, but a temperature profile chart prepared by Hank contrasts the temperatures at his houses at a particular hour.

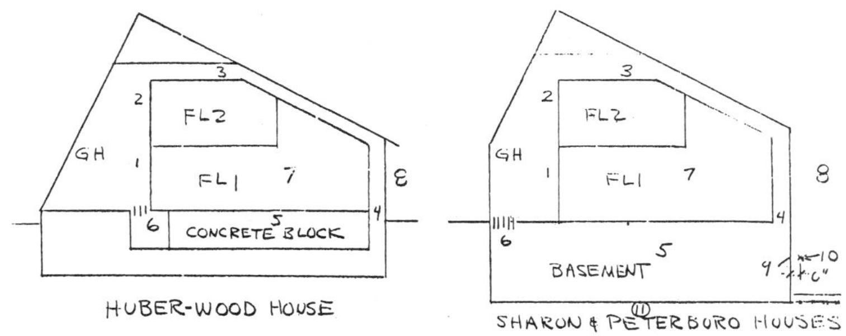

TEMPERATURE PROFILE °F 3-23-80 1 PM

HOUSE - CONDITIONS / TEMP POINT	HUBER-WOOD NO VENTING GRAVITY CONVECTION NO SUPP. HT	SHARON VENTING AT TOP GRAVITY CONVECTION NO SUPP. HT.	PETERBORO NO VENTING FORCED CONVECTION ~2000-3000 CFM NO SUPP. HT.
1	100	92	82
2	120	100	—
3	—	—	100
4	78	70	85
5	—	57	74
6	60	64	74
7	73	72	76
8	46	48	50
9	—	50	60
10	—	50	56
11	—	47	52
12	—	NO FLOW	43

NOTES PT11 ~ 10" BELOW BASEMENT SLAB
PT12 WATER TEMP OF PERIMETER DRAIN FLOW
PETERBORO HOUSE HAS 2" BEADBOARD UNDER SLAB, NO FLOW IN BASEMENT DRAIN

Comments by Hank

The greenhouse peak temperature at the Huber-Wood house is substantially higher than the other two houses, probably a combination of better outer shell insulation (R30-40 compared to R19) and the increased air flow resistance of the concrete block storage system. The Peterboro house (forced convection) shows substantially higher (74°F) basement temperature than the Sharon house (57°F). Most significantly, the surface (60°F) and 6" depth temperature (56°F) in the Peterboro house are substantially higher than those of the Sharon house (50°F). Much more heat is being delivered to the basement storage. Four weeks prior, when forced convection was installed, the Peterboro house thermal mass temperature was 3°F cooler than the Sharon house.

The graphs that follow are both of the Huber-Wood house. One is with forced convection and no insulation on the inner shell and the other (Feb. 5) is with gravity convection and full insulation on the inner shell.

February 5 was preceeded by several days of fairly good solation with overnight temperatures close to 0°F and mid-day temperatures of 18-25°F. No overnight supplemental heating was used. A one-hour fire was used to raise the inner house temperature from 53°F to 62°F. Bright sunshine took over at about 9:00 a.m. and continued to raise the indoor temperature to 72°F. Morning envelope temperatures ranged from 48°F in the north cavity to 42°F at the bottom of the greenhouse. The lowest temperature monitored in the greenhouse during the past winter was 40°F. A Delta T of 30°F occurs between the greenhouse peak and the concrete block storage. An additional 19°F drop occurs through the block.

Forty percent (19/(30+19)x100) of the loop drop occurs through the block. This is called by Jim Berk the earth (block) transfer differential (ETD). Of the net solar energy gained by the house, forty percent is being stored in the sub-floor thermal mass. This is substantially higher than any other monitored envelope house. The best performing monitored house, the Burns house, shows an ETD of only 5°F. It is difficult, on the basis of this data, to directly compare the Burns house and the Huber-Wood house. The Burns house is a better insulated structure, tighter built, considerably smaller, lower surface to volume ratio and doesn't have a three story silo on the west end.

The graph of April 6, 1979 is of interest primarily in showing the ability of forced convection to control peak greenhouse temperatures. A mid-day ETD of 16°F compares well to 19°F one year later with gravity convection.

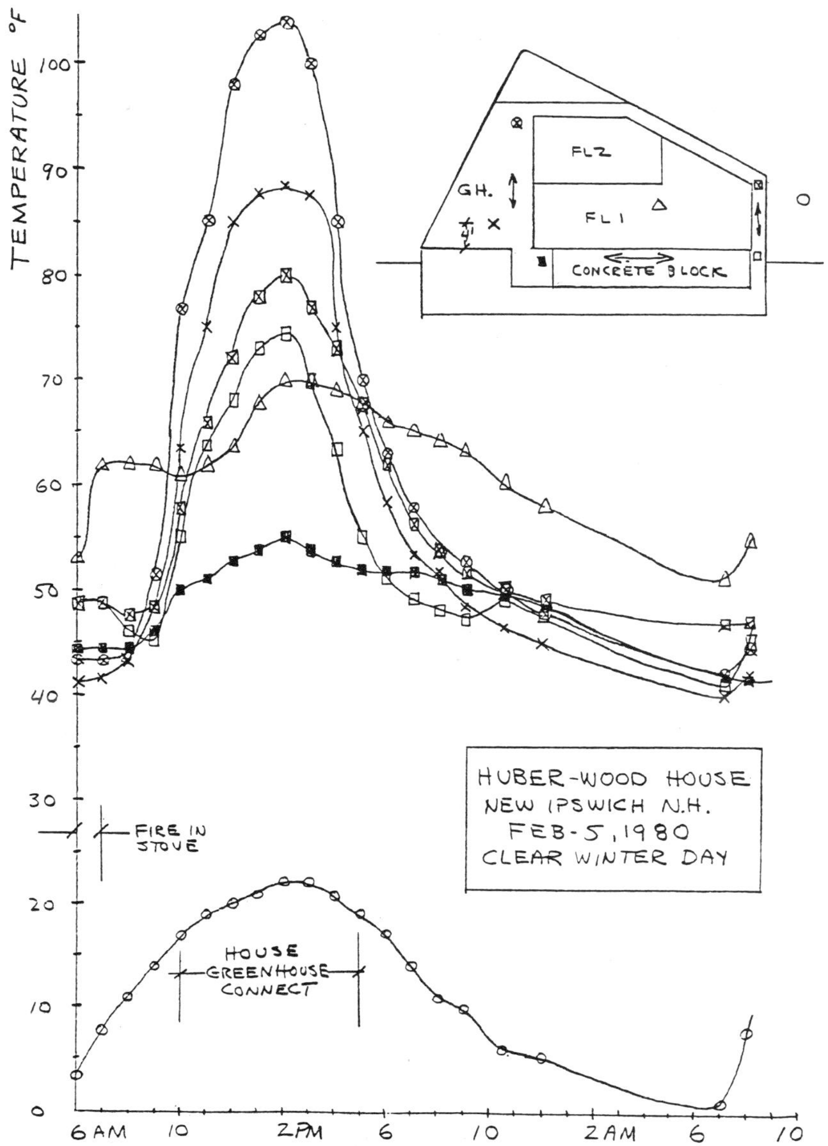
TEMPERATURE °F
100
90
80
70
60
50
40
30
20
10
0
6AM
10
2PM
6
10
2AM
6
10
GH.
FL2
FL1
CONCRETE BLOCK
FIRE IN STOVE
HOUSE GREENHOUSE CONNECT
HUBER-WOOD HOUSE
NEW IPSWICH N.H.
FEB-5, 1980
CLEAR WINTER DAY

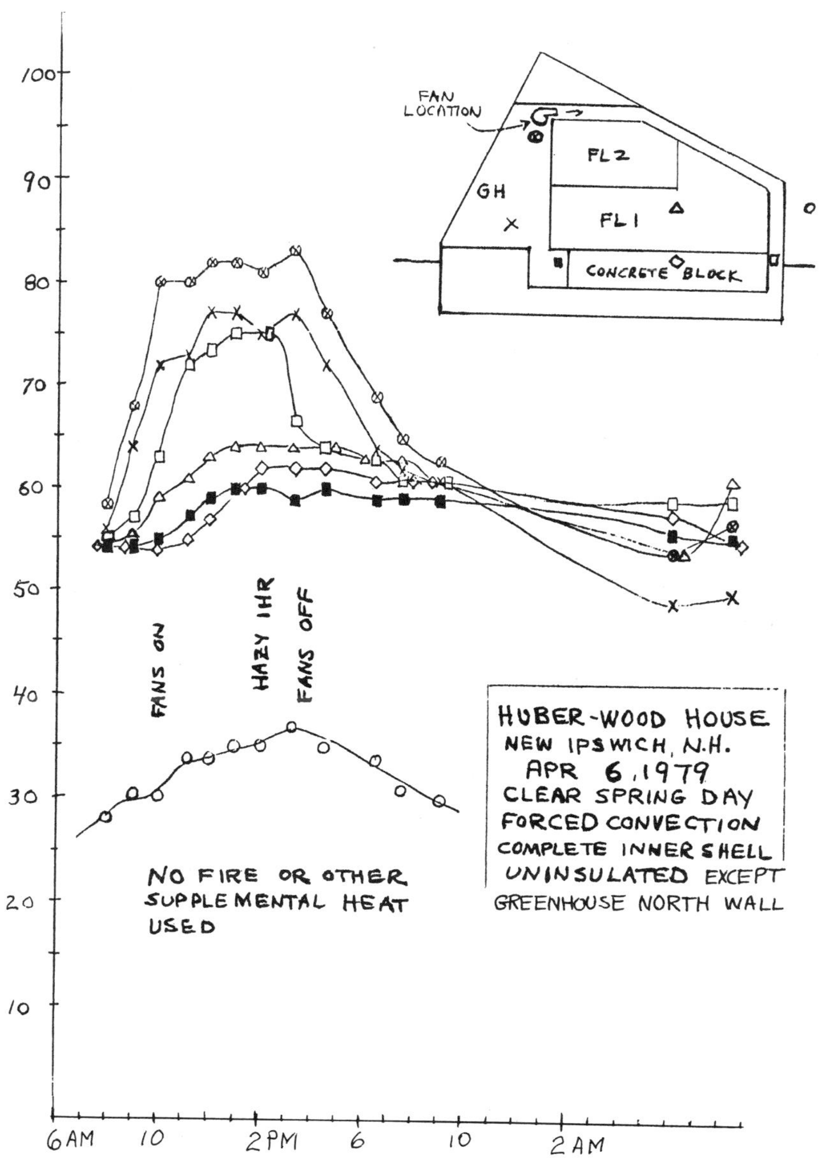
100
90
80
70
60
50
40
30
20
10
6AM
10
2PM
6
10
2AM
FAN LOCATION
GH
FL2
FL1
CONCRETE BLOCK
FANS ON
HAZY 1HR
FANS OFF
NO FIRE OR OTHER SUPPLEMENTAL HEAT USED
HUBER-WOOD HOUSE
NEW IPSWICH, N.H.
APR 6, 1979
CLEAR SPRING DAY
FORCED CONVECTION
COMPLETE INNER SHELL
UNINSULATED EXCEPT
GREENHOUSE NORTH WALL

8. SIMULATIONS by Doug Clayton

The performance of envelope houses can be predicted, analyzed and optimized by performing energy balance calculations, assigning conductances and heat capacities to all portions of the building interacting in the particular case being considered, and solving for temperatures or energy transfers.

For example, a simplified energy balance for the envelope might be expressed:

Heat loss envelope to ambient = solar gain envelope ± heat released/absorbed by the masses ± heat gain/loss from/to inner building + internal gains of envelope

and for the inner building:

Heat loss inner bldg. to ambient = solar gain inner bldg. ± heat released/absorbed by inner mass ± heat gain/loss from/to envelope + internal gains

Unfortunately for such calculations to be useful predictors of real world dynamics they must be performed at frequent intervals. Solving for a succession of temperatures and/or energy transfers, each calculations making use of preceeding solutions, would be impossibly time consuming without the aid of a mathematical genius or computer.

We will look at two computer models presented at the March 15th conference, the TEANET and Norman Saunders'.

In the fall of 1978 Community Builders purchased a Texas Instruments TI-59 Programmable Calculator and Total Environmental Action's TEANET Program, to use in evaluating the envelope's effectiveness. TEANET is a seven node thermal network algorithm developed by TEA to simulate the performance of passive buildings. The TEANET Program performs an energy balance calculation, solving for temperatures at seven nodes and auxiliary energy use on an hourly basis. The program internally generates ambient temperature and solation sine waves based on the daily weather information provided it.

Compared to other programs with 100 or more nodes (e.g. Norman Saunders' that follows), the TEANET is rather unsophisticated, but it may be adequate if the proper assumptions are made and the thermal network correctly conceptualized. We had assumed, after a week or two of working with the program that the major and perhaps fatal limitation of TEANET in modeling the envelope was its inability to simulate changing convection flow rates and direction. Attempts were made to overcome this limitation by stopping the program whenever we guessed there would be a significant change in the volume and/or direction of flow (e.g. at sunrise, sunset).

We never assumed flows greater than 4 cfm per sf aperture (or velocities above the level of human perception). All this assuming was very scary, but in retrospect it appears that we were in the ballpark. With the new awarenesses expressed elsewhere in this book regarding the interplay of envelope configurations, temperatures, densities, pressures and velocities, appropriate flow rates for the TEANET may be assumed (see page 7.10).

We first applied the TEANET Program to the Sant Bani School building, an envelope we were designing in the fall of 1978. It was very exciting to read the first tapes spilling out of the printer and see the envelope "work."

The model stored and retrieved heat. Just as anticipated, the envelope experienced much wider temperature swings than the inner building. Different glazing/insulation schemes were tried to reduce these fluctuations in envelope temperature, but probably what we were missing was the importance of the interplay of the mass of the building's structure.

The model pointed out the need to get some BTU's into the inner building first thing in the morning, and deal with the excess by afternoon. TEANET also told us we had to do something about the infiltration associated with the normal traffic of 150 young bodies in and out.

All this was very time consuming, but by the time the Burns house was ready for TEANET's consideration (March 1979), I was familiar with the program, and developing the schematic and plugging in the values was a relatively simple task.

The envelope portion of the Burns house was conceptualized as two nodes. One was the aperture and roof portion with all the gain, most of the heat loss to the outside (ambient) and little mass (first ½" of sheetrock or wood surfaces). The other was the bermed north plenum and basement/crawl space. This second envelope node was also considered low mass and the earth/concrete masses were assigned nodes to give and take heat to and from it. Both envelope nodes gave and took heat energy to and from the inner building.

It didn't matter to the program which way the air might be flowing in the loop since it was simply a conductance between the two nodes. For the five-day simulation, December 17th to the 21st graphed here, the program ran with a continuous 20 fpm (460 cfm or 496 BTU/hr-^{o}F).

Air change values between the inner house and outside, envelope and outside and inner house envelope were guessed at by a synthesis of techniques (linear feet of crack, fan volume, etc.). That probably resulted in conductances greater than those actually occurring in the Burns' house (in this model 500 cfh inner house to outside, 2,900 cfh envelope to outside and 700 cfh inner house to envelope).

Other assumptions in this model:

Earth/concrete average surface conductance - 2 BTU/hr-^{o}F

Earth/concrete average conductance - 12 BTU/ft^{2}-hr-^{o}F-in

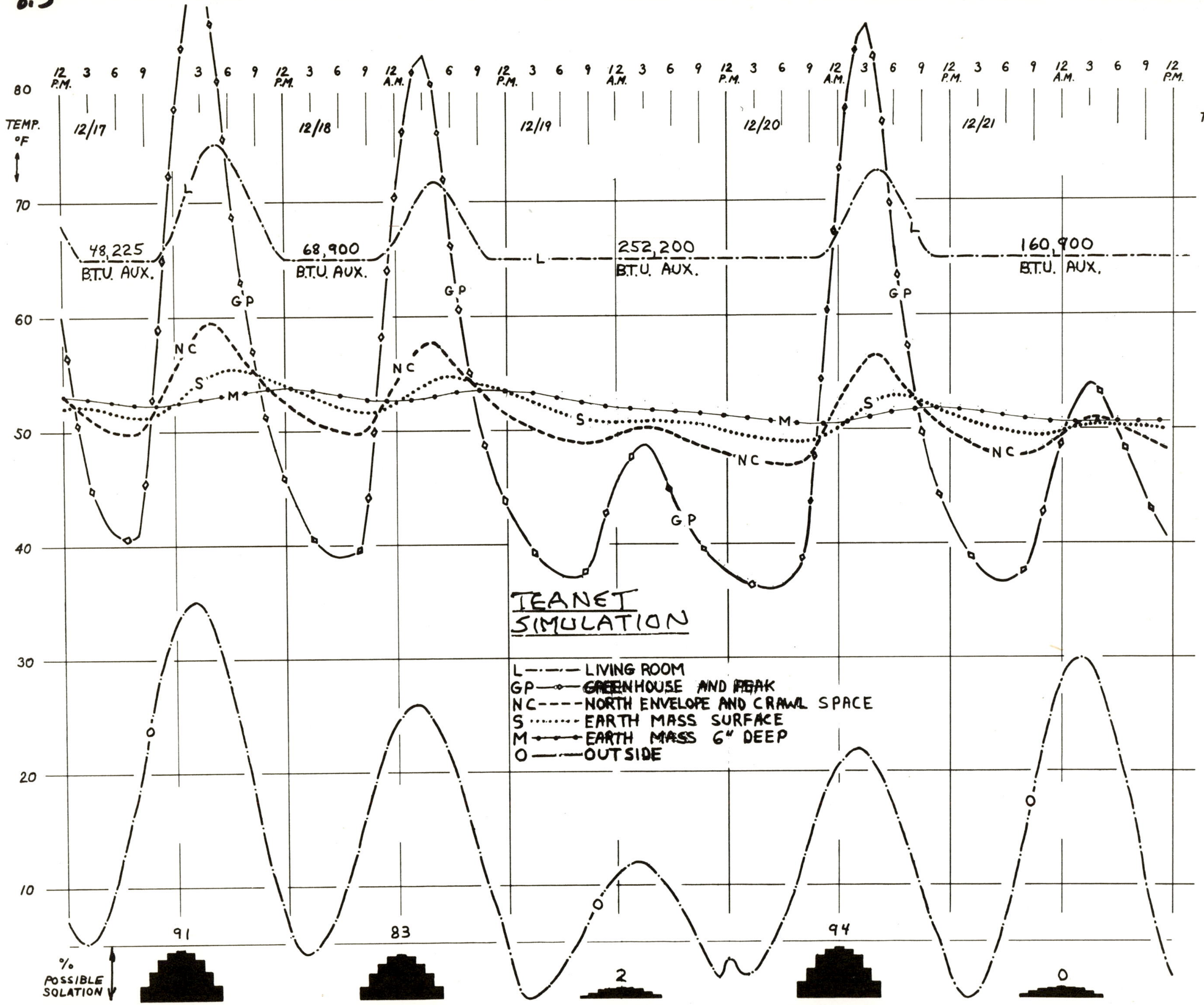
12 P.M. 3 6 9 3 6 9 12 P.M. 3 6 9 12 A.M. 6 9 12 P.M. 3 6 9 12 A.M. 3 6 9 12 P.M. 3 6 9 12 A.M. 3 6 9 12 P.M. 3 6 9 12 A.M. 3 6 9 12 P.M.
12/17
12/18
12/19
12/20
12/21
TEMP. °F
80
70
60
50
40
30
20
10
48,225 B.T.U. AUX.
68,900 B.T.U. AUX.
252,200 B.T.U. AUX.
160,900 B.T.U. AUX.
L
GP
NC
S
M
TEANET SIMULATION
L LIVING ROOM
GP GREENHOUSE AND PEAK
NC NORTH ENVELOPE AND CRAWL SPACE
S EARTH MASS SURFACE
M EARTH MASS 6" DEEP
O OUTSIDE
91
83
2
94
0
% POSSIBLE SOLATION

Earth/concrete capacitance - 35 BTU/ft^3-^{o}F

Average internal gains - 800 BTU/hr

Other conductances and capacitances were calculated using standard ASHRAE values and techniques.

In retrospect the model seems to be a good one. The most notable discrepancy between the simulation and recorded data seems to be the low temperatures reached in the simulated greenhouse. Either the envelope is more conserving of energy than calculated or there are some BTU's sneaking in from somewhere, perhaps from the mass of the structure or deep earth that was left out of the model. The model seems to have a poorer coupling between its envelope and earth/concrete masses.

The auxiliary BTU load calculated by TEANET to maintain 65°F is roughly equivalent to 53 pounds of firewood, or more than twice what the Burns house actually used during the modelled period. This auxiliary load would have dropped significantly if a less conservative internal gain value had been assumed.

Norman Saunders presented his new computer program, specifically designed to simulate envelope houses at the March 15th and 16th conference. The following is paraphrased from his presentation:

I've spent two months designing and building a computer program to model double envelope houses starting from basic physical principles. Last Thursday afternoon I finally got the bugs out to the point where I felt I really could use it. So Friday morning I put in the data for the Burns house and this is the first run (holding up five feet of computer print-out)--this is the start of it. The machine computes every five minutes, in SI (System/International) and prints out in English every hour.

This is the week of Dec. 17th through part of the 20th (see graph). The bottom line there is 40°F, 50°, 60°, 70° and so on. I did not have the solation set correctly and the house sort of slid down in temperature, but if you look at the details there and compare it with the measured data, you'll find they're very close, with a couple of exceptions. When the oven went on, the temperature rise was less than the actual oven, because I apparently underestimated the power used by the oven. And you'll notice that the temperature jumps on these two days when wood was burned around 2 or 3.

The big problem in the simulation is the things we don't know. I had to guess at infiltration, heat of occupancy, oven heat, etc. What I did was divide the structure up into about a dozen different surfaces and each of those surfaces then was divided up into about a dozen different layers. So that the time delays in real life are there. You'll notice that the north temperature comes up after the greenhouse does, and if you look at the individual layers, which I don't have any plots of, you'll see the delay in heat working through from one layer to the next. The greenhouse is perhaps reaching a slightly higher temperature than it should, but the detail is there.

The crawl space and deep earth I think I've got aligned about right. I should have had a better vertical alignment target on them. And you see the computed values show the same cycling above and below, that we have on the

8.5

80325 Performance of the Burns House for 79D17 to 79D21
As first calculated by DOUBLE ENVELOPE program
co Norman B. Saunders, P.E.   15 Ellis SCde   Weston MA 02193

Double Envvelope

Solid line is maximum air velocity at five feet per minute per division.
Living space, Greenhouse, North, Crawl space, Mass of earth same line
symbols as on plot of experimental data.

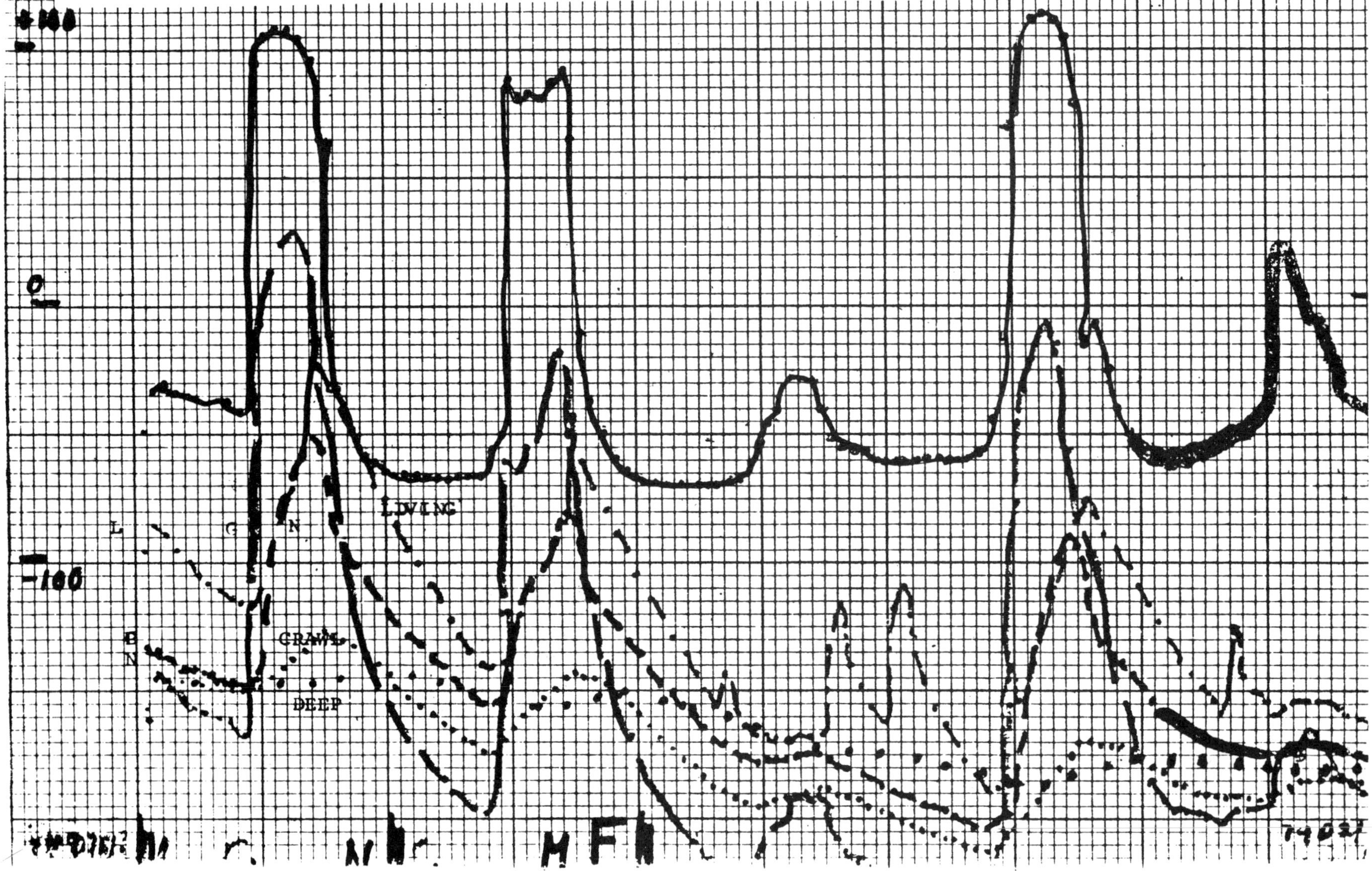

plot from measurements.

Then one thing which you don't have on the measured data, but which the data processor does turn out, and this may be the most interesting of all is flow velocity. The plotted value is the maximum in the vena contracta (area of maximum velocity slightly downstream from constriction) at the lower north wall. The appropriate lesser values were used in the calculations for each surface. Scale is on the left of the graph, +100 to -100 feet per minute. The flow reverses about an hour or two after the sun comes up, and you notice it seemed to reverse quite rapidly. In the afternoon there is a burble when the flow is about to reverse. It shows up more in the temperature plots than it does in the flow plots. And you'll notice that the one day there's very little sun, the flow sort of slowed down, but it never did go in the forward direction.

Here at midnight between the 16th and 17th are the initial conditions which don't match well. You see with over 100 points, 100 nodes in the program, setting them all up at once is a pain in the neck. So I simply approximate the initial conditions so the first hour's a little off. But within an hour it's pretty well settled in.

So we come along here, and get up to around 9 am at which point the flow, having been -40 feet per minute or something, jumps up fairly quickly in the course of a couple of hours, up to nearly 100 fpm in the "forward" heat collecting direction, up in the greenhouse, down in the north. Then in the afternoon it hangs on, around 5:00 or so it comes down, then reverses. This was a cold night, so it reversed fairly heavily. The next day the 18th, the same sort of thing happened, but it wasn't quite such a bright clear day, so there's a little hole in the flow, it's slowed down a little bit. On the 19th, a cloudy day, the flow reduced in velocity, but didn't quite get around to reversing. And then we got another sunny day, the 20th, where she went up nicely.

Now as to the magnitude of the flow there, I have run some tests on the sensitivity to the flow resistance. This is one of the things I really didn't know, what the flow resistance was. According to my calculations, its primarily in the bend at the north wall crawlspace place, that being the point of greatest velocity and the momentum is mostly lost. Essentially you lose the dynamic pressure head there. I made computations with various flow resistance and I found that the temperatures calculated were not at all critically dependent upon the flow. I found that I could vary the flow restriction by 4:1 with almost no change in the temperatures that resulted from the computation. And for the plotted data I used a slightly greater restriction to flow than I calculated from the details of construction of the house.

Remember that the flow is a square law proposition and so it's rather insensitive. The pressure you've got to work with remains relatively constant so we're relatively insensitive to flow, and pressure differences, things of this sort. You have a temperature difference between the greenhouse and the north wall. And that temperature difference results in a density difference. That density difference gives you a pressure difference. That pressure difference then forces air around the loop, up to the point at which the pressure drop is equal to the pressure you had driving, and the flow resistance seems to be primarily in the dynamic pressure drop. That is, if the air that's moving has momentum, and if it changes direction suddenly, as we do in most of these houses, you lose that momentum, you lose that energy.

The quantity of heat available is relatively constant so velocity times temperature head is relatively constant. For a fixed flow resistance the pressure drop varies as the square of the velocity. Put these together and in the double envelope house the loop velocity varies roughly as the cube root of the resistance to fluid flow.

As far as temperature goes, the thing that we see, of course, is that the air reaches its maximum temperature as soon as it's left the last sunlit area, and from then on around the loop it gradually drops in temperature, so by the time we get down to the crawl space there's very little of the temperature rise left.

If you make the turns easier you get a higher rate of flow. With a higher rate of flow you get better heat transfer between the air and the surface. The better heat transfer means that their temperature rises are less, and the temperature fall is less, so there's less density difference to drive the air, and since there's less pressure to drive the air, why this cuts down.

In my own analysis, which I do in my head, I take into account all of these sundry little loops and all. But in the data processor, it would just take too much store and too much time, to take all that into account. This is a very simplified analysis of the thing. I know it's got 20 times the number of nodes that the TEANET does, but nevertheless when you're trying to actually model the entire thing, it's a very simplified analysis. And I took the loop as being a single loop. I did write a paper yesterday, on the within-loop flow; I observed them to be viscous. But around the loop I considered it to be turbulent. This is the around the loop flow. I do not have a plot of the flows within the spaces themselves. In reality, both can go on at once, and it can be very complicated. Actually, Berk's paper is about 90% right in describing what goes on. If you want details, look at that.

I ran the program with the loop completely shut off, no air flow in it whatsoever. That house temperature dropped like a rock. I don't believe it, but that's what the data processor told me. Of course, I've spent a lifetime building and designing data processors and I only use it when I have to, I don't trust them. So then, since that happened, I said, "well, maybe its because I left in the infiltration." So I left out the infiltration and let it run, that's the way it ran last night. Well, this morning after about ten days simulation time the house was down to around 30°F, so I said, "well, all right, so shutting off the loop has really done something." When I put the loop back in the house started working. For Dec. 17th, with the loop shut off, the temperature dropped 15°F compared with full simulation. Without the flow to move the energy gained it's lost right back out the glass.

The program is available on a consulting basis. It takes about a day to analyze the proposed design for its major elements, determine the values for them and key them into the machine. Another day is allowed for alterations in the design to see how it can be improved.

9. SINTEF

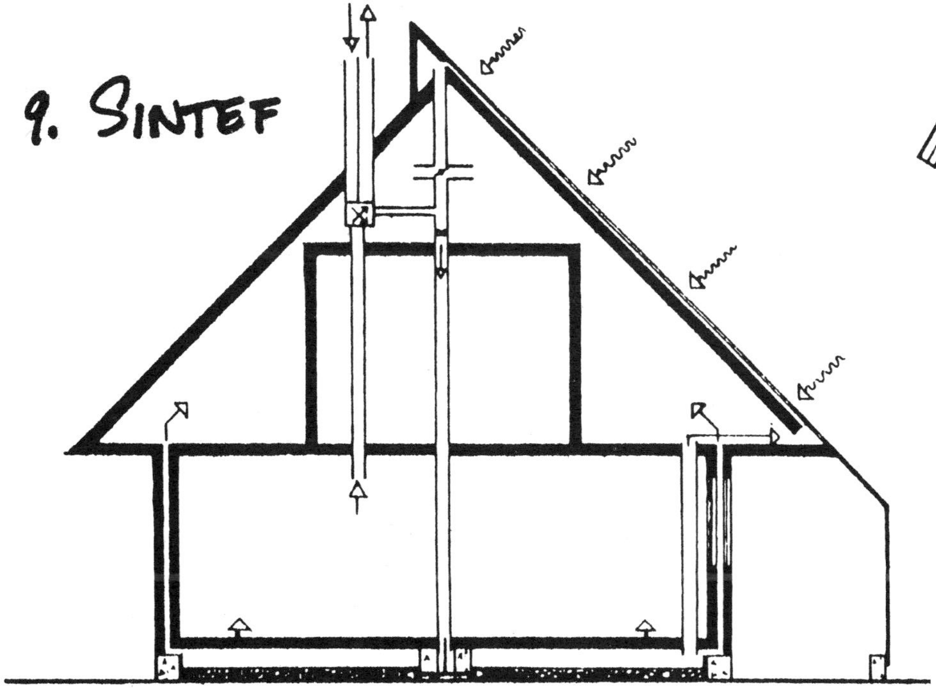

At the Foundation for Scientific and Industrial Research at the University of Trondheim in Norway, Johannes Gunnarshaug and associates have been pioneering double-shell systems for several years. We've translated a 1979 report.

This summer's SINTEF program expects to demonstrate economical, completely solar heated houses at the Arctic Circle, through the use of a small heat pump to boost earthstore temperatures as low as 40°F. Water in plastic tubing buried six feet deep in the earth under the building will carry heat to and from the earth store. Summer water temperatures will be up to 100°F.

SOLAR ENERGY FOR THE HEATING OF HOMES

RESEARCH BUILDING AT DANSKERUD IN ÅS, NORWAY

EXCERPTS FROM PRELIMINARY REPORT

BY Torbjørn Teitgård and Johannes Gunnarshaug
December 1979
Translated by Robert Schongalla

THE BUILDING

The building is built with 1½ stories without a cellar. The first floor is constructed of bearing "Multi-elements". (Gypsotet Multi-element is a sandwich-element with outer sides of gypsum sheets and an inner layer of polyurethane foam). The living space in the second floor is constructed as a box of "Multi-elements" placed inside the roof rafters.

The outer walls are made of double "multi-element."

9.2

An air space between these two elements is constructed with the help of vertical laths. The windows are set in as separate parts. In the inner "Multi-element" there is a sealed, inwards-opening window, and in the outer element there stood originally a single fixed window which was changed to be a double window during the fall of 1979, because of condensation and ice build up on the outer window.

Special air vents are placed in individual places near the side of the windows. The air vents are caulked on the air space side. The doors to the balcony and the winter garden can also be used for airing.

The inner wall and ceiling is made up of single "Multi-elements". The ceiling elements are fastened to the roof rafters. The roof rafters have a pitch of 45°. "Multi-elements" are laid down on the rafters as outer insulation. The inner insulation part of the second floor is also single "Multi-elements".

The roof on the north side is covered with aluminum in the usual way, but the roof on the south side is equipped with a 45.5m^2 solar collector, in other words with an absorber and one layer of glass over the "Multi-elements". A winter garden is constructed as a continuation of the solar collector in such a way that the glass layer continues down and is widened out into a half climatized living space.

SOLAR HEAT SYSTEM

Mode of working of solar heat system

Air is drawn in from the air space in the solar collector. In the end of the collector there is a collection duct and the air is drawn farther by a fan of 120 W via the vertical main duct to the heat store under the house. The main duct has a diameter of 300 mm. From there the air flows out in a distribution duct and is forced out to both sides of the heat store which is made up of a layer of stones. After having passed the store, the air goes out into a crawl space under the floor and then flows up into the air space between the walls.

In case there is no need for heat in the living spaces, then air is not wanted up in the air space. One can then open the dampers of two bypass ducts which are 20 x 40 cm and most of the air will flow out of these. The dampers were originally equipped with a damper motor and thermostatic control. Experience has since shown that they are only needed at the beginning and end of the heating season and this change is now being done manually. When the air which flows through the solar collector is not furnishing heat, a damper opens automatically which permits air to be drawn directly from the loft rooms to the main canal. This is controlled by heat sensors located in both ends of the solar panel.

Air circulation in the solar heating loop is 1500 m^3/h.

The solar collector

The solar collector has a surface area of 45.5 m^2. The absorber is made of aluminum sheet painted black. It is cooled by the air which flows between the absorber and the underlying outer roof. Reradiation of heat is prevented by a layer of glass of the kind that is used in greenhouses. The glass is placed with about 1 cm overlapping. The space between the glass and the absorber is vented into the winter garden. The winter garden's roof is simply a continuation of the glass layer of the solar collector. The "Multi-element" sheets which form the building's outer roof have in addition the task of insulating the solar panel on its underside. Caulking of the solar collector is done with silicon.

The heat store

The solar heat is stored under the house in a 16 cm thick layer of stone spread out over the whole foundation area of 80 m^2. In all, there is about 13 m^3 of round river stones with a diameter of between 5-10 cm. In order to prevent air leakages and moisture penetration, the store, both over and underneath, is covered with a coated plastic foil. The space between the stone store and bedrock is backfilled with sand and local material which are well packed in order to permit good heat transfer downwards. In this way the mountain bedrock under the house can be used as a heat store. The foundation wall is insulated under the back side on the outside with isopor. On the east side the insulation goes down 1m. On the other three sides, it goes down to bedrock and from there about 0.5m out as flank-insulation. Between the heat store and the floor of the house there is a crawl space.

The floor is insulated with 5 cm heavy mineral wool. The stone store is not insulated on its top side, and a certain direct heat transfer will occur between the store and the living space.

Ventilation

The system is constructed such that 100 m^3/h fresh air can be supplied to the circulating air in the solar heat loop. At the same time, an equal volume of air flows into the living space from the wall space via vents in the inner wall. A heat recuperator draws used living space air through outlets in the kitchen, washroom and bathroom and preheats the fresh air as is being drawn in. This mechanical ventilation is, however, used for only short periods as the occupants have found it unnecessary to have it on continually.

Preheating of household hot water

For the preheating of hot water there is a 150 m long pipe of cross-bound polyethelene in the stone store. The pipe holds about 60 l of water which is preheated before it reaches the hot water tank. By switching a valve in the kitchen, it is possible to tap warm water directly from the pipe in the stone store if desired. In the case that need demands it, water can also be taken directly from the hot water tank.

Temperatures and energy flow in the solar heat system

As far as temperatures in the solar heating system are concerned, the first complete data are from March 1979. The reason for this is certain problems in the beginning. Figure 5 shows the quantity of energy taken up by the solar panel from the 11th to the 20th of March and the distribution of this to the heat store, wall space and heat loss. According to the measurements, there was during this period an average absorption of 70 kWh/day in the solar panel. This corresponds to an efficiency of 35-40%. Eighteen percent of the energy went into the wall space and 54% was carried to the heat store. In the case that the stone store had been insulated, the temperature in it would have increased by many degrees each day. But very moderate temperature variations were measured just under the heat store (see Figure 5). This means that the energy was quickly transported down into the mountain bedrock.

If the stone store was insulated, heat to cover about three winter days' need would be able to be stored. By also making use of the ground, the storage volume is increased at the same time since the heat loss is greater.

Temperature recordings at depths of 1.5m and 3m confirm that heat is transported downwards. Periods of a great deal of sunshine create "ripples" in the temperature curve at a depth of 3 m also. Heat flows always in the direction from high to low temperature. Heat flows upwards only during December and January when, consequently, the long-term store of solar energy is used.

EXPERIENCE BASED ON ACTUAL OPERATION

The solar panel - During heavy rain quite a bit of water leaked in the junctions between the glass plates in the solar panel. In one case, water leaked in under the aluminum plates and from there, into the loft via the junctions between the "Multi-elements." After the glass junctions were caulked at the joints with silicon in the fall of 1979, the problems seemed to be fewer. Condensation on the inside of the glass has generally not occurred when the space under the glass is vented to the winter garden.

The fan - Because of noise from the fan motor, the occupants had to turn off the solar heat system during the night.

The controls - The automatic regulating of the by-pass dampers which has been mentioned was not used and will be taken away. The goal with this was to lead the circulating air by the side of the wall space if the temperature in the living space was too high. Uncomfortable overheating has not occurred; possibly the air flow in the wall space works in a cooling fashion in such cases. The dampers are now regulated manually at the beginning and end of the firing season.

In the summertime, there is no reason to have the solar system working when there is no solar energy yield. There will, therefore, be installed controls which will cut in during the summertime to start or stop the fan, depending on the temperature in the solar panel.

Condensation on the windows - There were real problems with condensation and formation of ice on the outer window pane. A possible reason is that moist air from inside the house penetrated the wall space. According to the original plans, the circulating air would be ventilated with fresh air via the heat recuperator. This, as mentioned, was used only during short periods. The outer window panes were made into double ones during the summer of 1979 and the problem was thus solved. No moisture damage has been observed in the walls.

DISCUSSION AND CONCLUSIONS

The integrated character of the solar heat system makes it impossible to measure useful solar heat directly. But it can be asserted that the solar heated building has considerably lower heating needs than buildings with corresponding theoretical insulation standards at the Danskerud grounds. The heating need of the house for 1978-79 was 5,900 kWh delivered by electric heaters, plus 900 kWh for the operation of the fan, while four houses with corresponding insulation standards used from 14,000 to 17,000 kWh. This large difference can well enough not be in its entirity attributed to the heat produced by the solar heat system. A considerable part of it can be traced to fewer air changes in the solar heated house. Tests of the "tightness" of the house using the Swedish "high/extra pressure method" show this. But it is not possible to quantify the differences in real air changes on the basis of these measurements.

The relatively low heat need is also due to special qualities of the double shell construction. This construction eliminates heat loss to the ground during the heating season and the air space between the inner and outer shell works as a heat recuperator for the ventilating air. The differences between the building and other buildings is not due to any real difference in average room temperatures. In the solar heated house, less electricity is used for lighting, cooking, hot water heating and electrical apparatus, not counting electrical heating, than in the other buildings. This leads to higher heating need.

This solar heated house works most fully in the middle of the winter when conventional solar heating systems produce the least amount of additional energy at our latitude. The house performed especially well during the recorded cold periods of the 1978-79 winter. During one week around the change in year, 1978-79, while the outside temperatures swung between minus 20 and 30 degrees, the electric heating consumed only 2.5 kW. To what degree this is due to heat from the earth under the house cannot be determined on the basis of the measurements, but temperatures measured in the ground indicate an upwards heat flow through December and January.

On the negative side of our experience, we should mention the preheating system for hot tapwater. This did not function satisfactorily. This was due to the heat flow down into the ground from the stone store during warm half of year which (heat flow) was higher than expected. Preheating of hot water ought to be done in the same way as in a regular air system with a heat exchanger in the collection duct right after the solar collector and with a large hot water storage tank as the heat store.

This solar heated building at Danskerud testing area is as far as we know the only solar heated double shelled house that is completed. Earlier, the principle was tested in an unlived-in research house in Trondheim. These two attempts have given good corresponding results with partly surprisingly positive outcomes which are not yet completely understood.

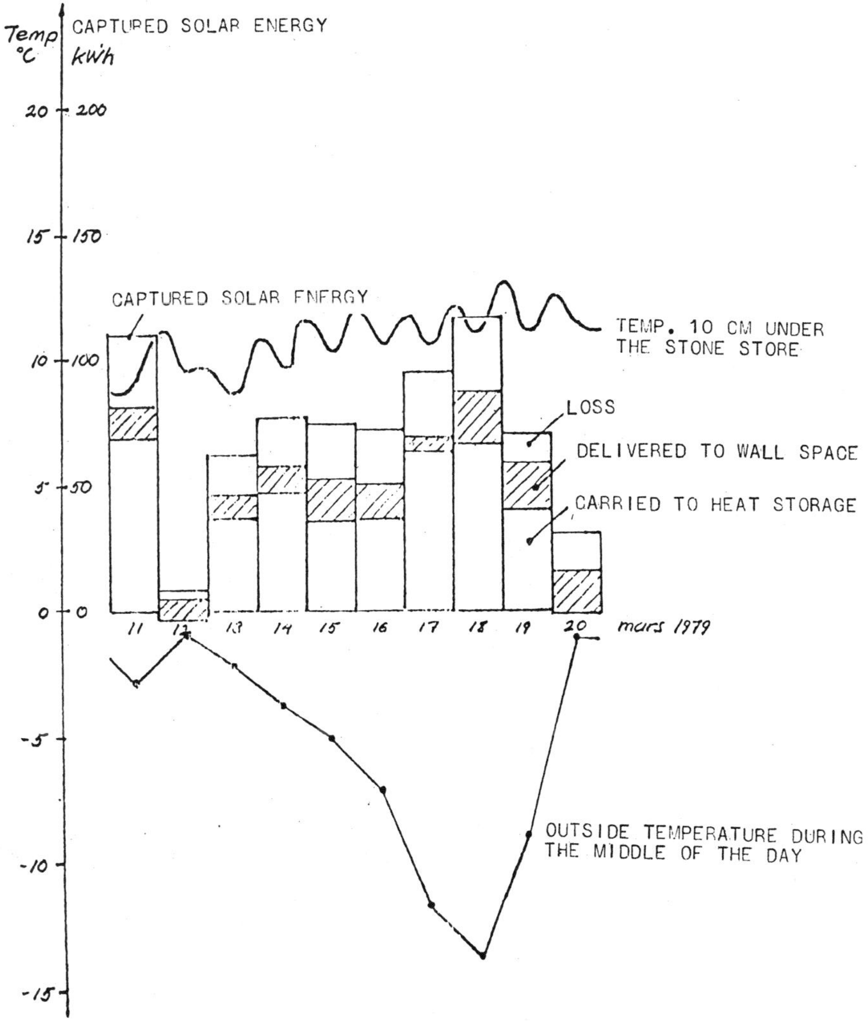

FIG. 5

SOLAR HEATING COMPONENT MARCH 11-20, 1979. CAPTURED SOLAR ENERGY AND DISTRIBUTION OF IT. TEMPERATURES UNDER THE STONE STORE AND OUTSIDE.

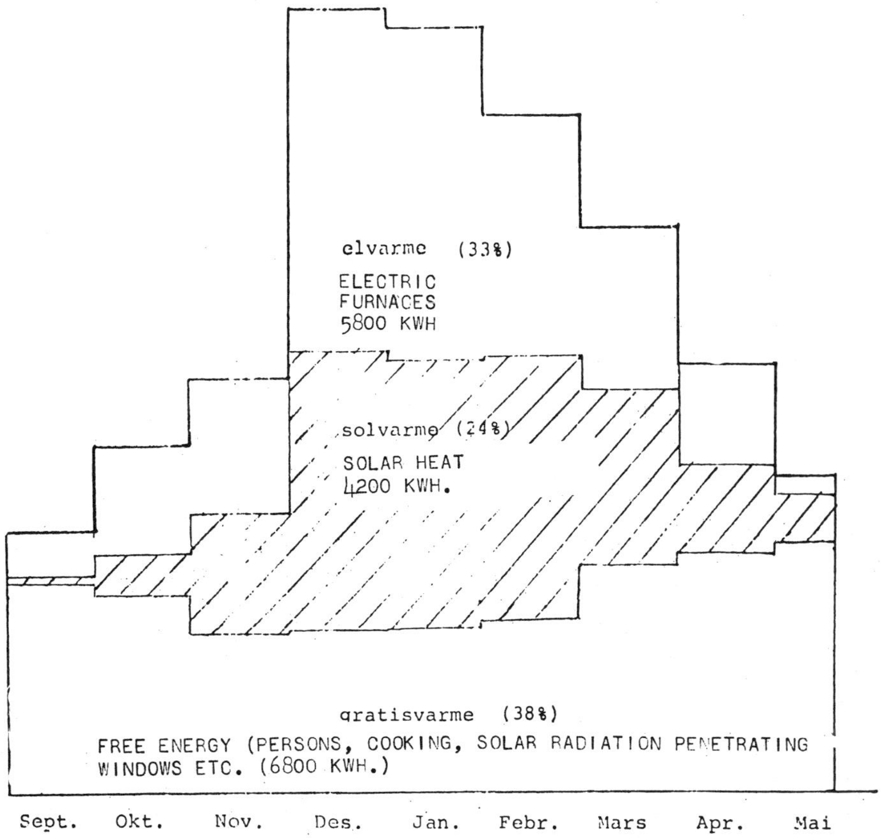

TOTAL TRANSMISSION AND VENTILATION LOSS FROM THE SOLAR BUILDING DURING THE 1978-1979 HEATING SEASON.

COVERING RESPECTIVELY: FREE HEAT, SOLAR HEAT, ELECTRICAL HEAT.

10. BACKGROUND: COMMUNITY BUILDERS

Community Builders was established in 1958 by an intentional community of families sharing their efforts and incomes. Having skills in the building trade, it was natural to offer these skills to provide an economic base for the community. After a short period of sub-contracting our services we turned to designing and building houses for sale, intending trouble-free quality at low cost.

By 1964 the community had reduced itself to the family of Lois and Don Booth, and Community Builders is now registered as a sole proprietorship but enjoys a very supportive relationship with other carpenters and associates--a team of skilled and caring people who enjoy working together and often discover better ways of doing whatever we're doing.

Shown in the photo opposite, left to right, are Bruce Pearl, Jeff Schadlick, Karl Moore, Peter Lawless, Don Booth, David Buzzell, Rodney Pearl, Bob Henninge, Kathy Chapman, Susan Shannon, Doug Clayton. Missing from the photo are John Andrews, Lois Booth, Seny Lake Jr., and Nancy Perry.

We planned this gambrel as a traditional-appearance, lower-cost envelope for people who wouldn't want the extra expense and care of a greenhouse. Instead there are plant spaces within the house in the deep-sill south windows.

Envelope airflow is at windows only, with the possibility of a small fan additionally blowing heated air from attic to basement down beside the chimney. All airflow spaces are gypsum-lined.

The roof-slope windows are not part of the envelope, except in an alternate version of the plan; they are Velux, with summer shade and inside thermal curtains.

Other plans shown are of the Sant Bani Ashram School (page 10-5) and of a sloping-site house (10-6 to 10-8) with entry in the north airflow and an insulated hot tub in the greenhouse.

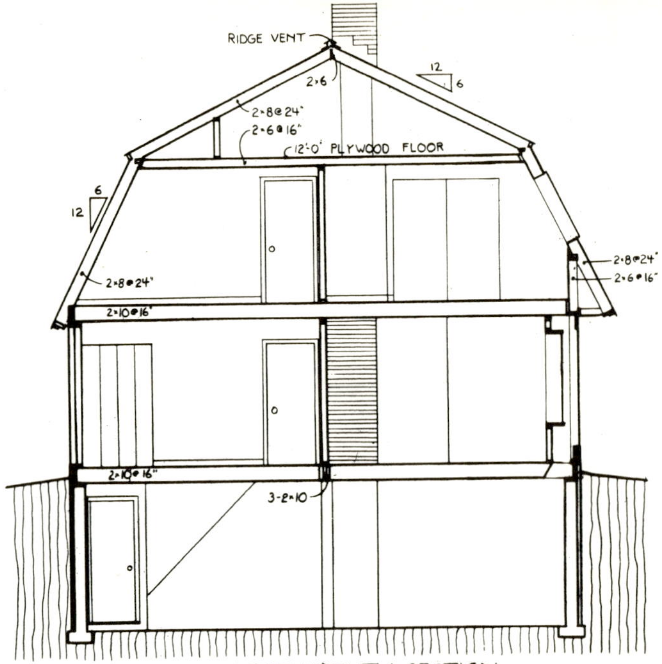

1/5 NORTH-SOUTH SECTION
LOOKING EAST

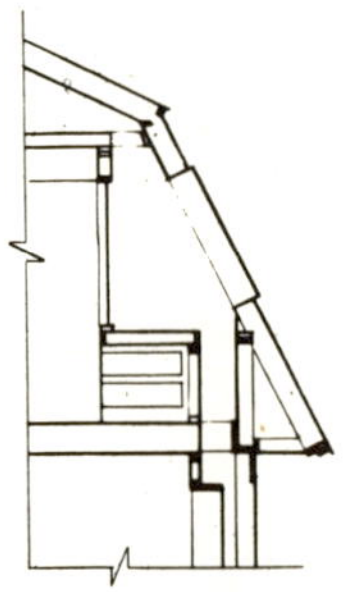

OPTION:
BUILT-IN DRAWERS UNDER INNER 4'x4' WINDOW

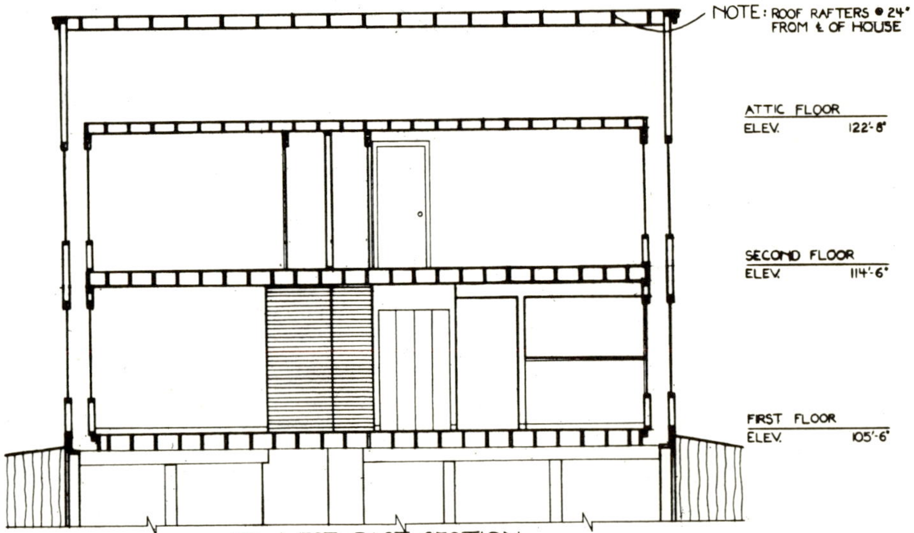

2/5 WEST-EAST SECTION
LOOKING NORTH

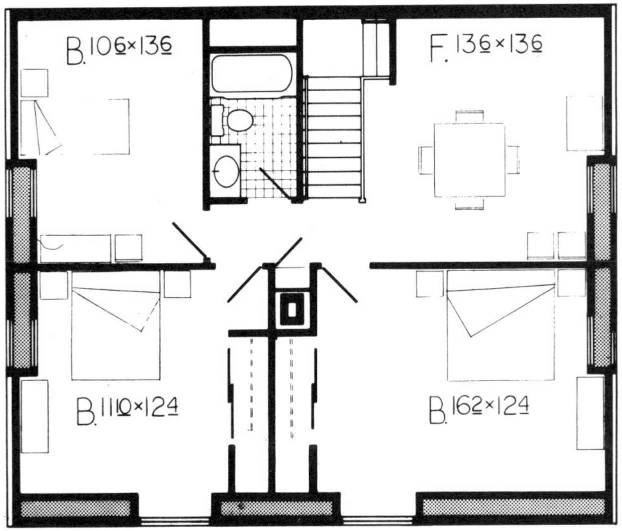

SECOND FLOOR

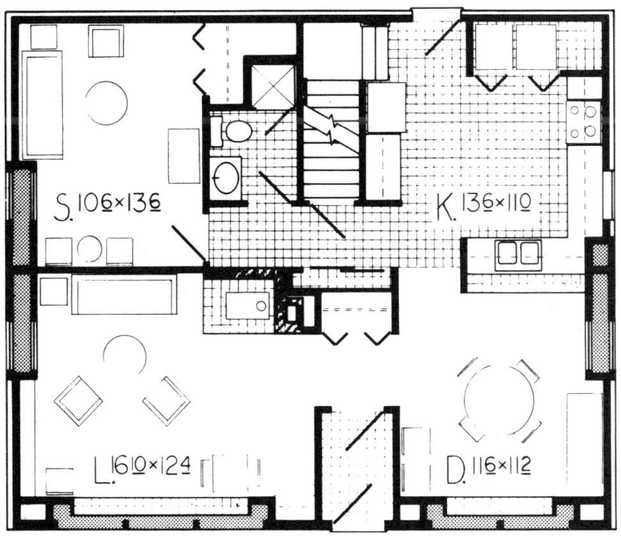

FIRST FLOOR

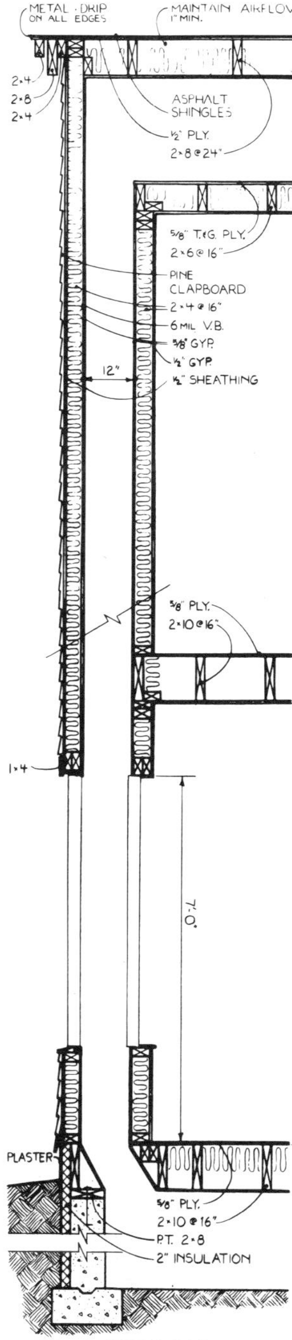

1/6 WEST WALL
AT DEN

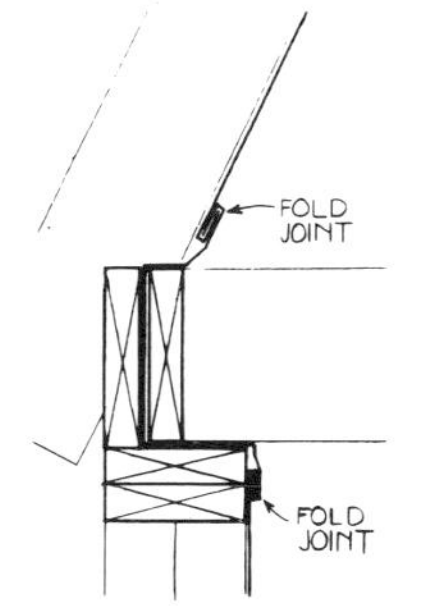

4/6 VAPOR BARRIER
DETAIL AT NORTH

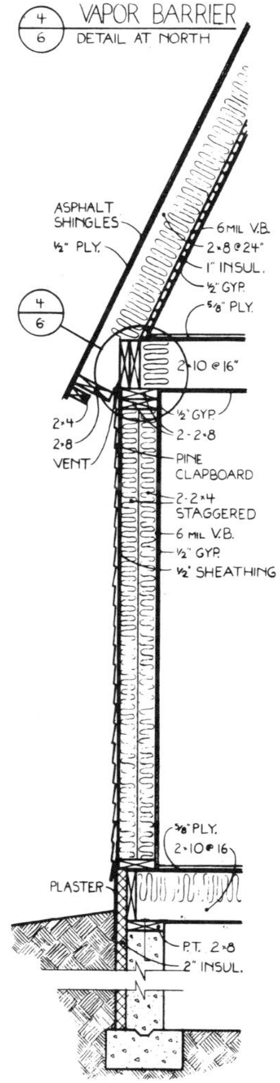

2/6 NORTH WALL
AT KITCHEN

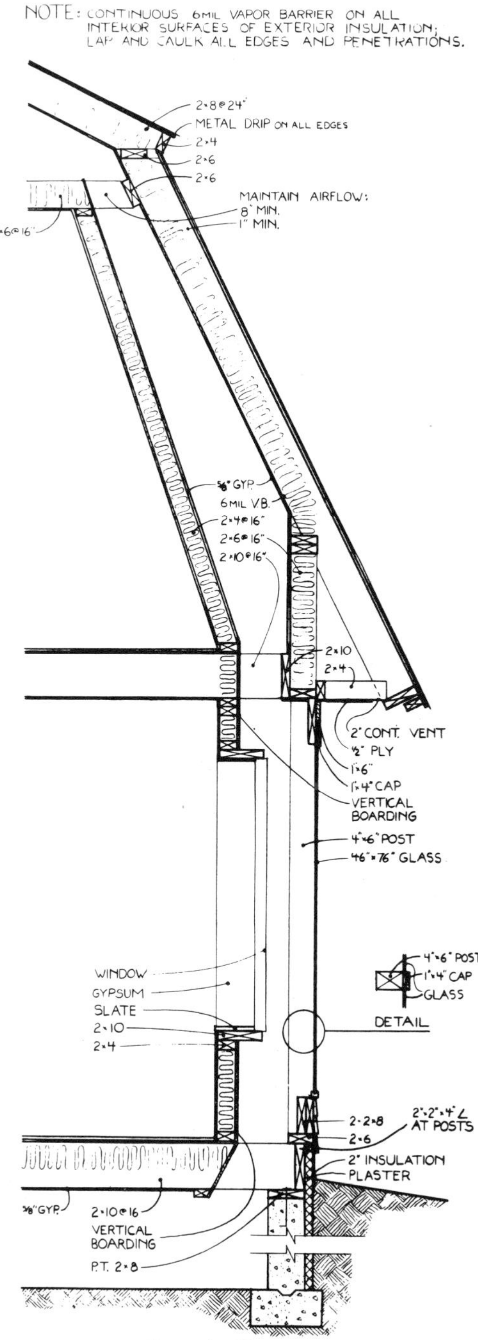

3/6 SOUTH WALL
AT LIVING ROOM

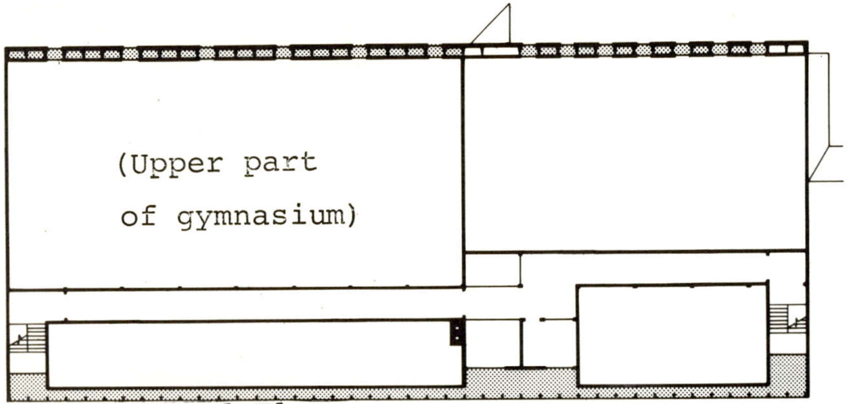

Second Floor

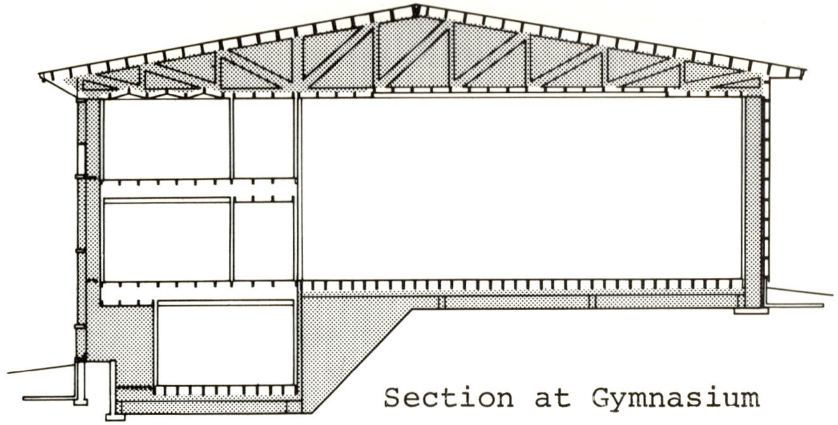

Section at Gymnasium

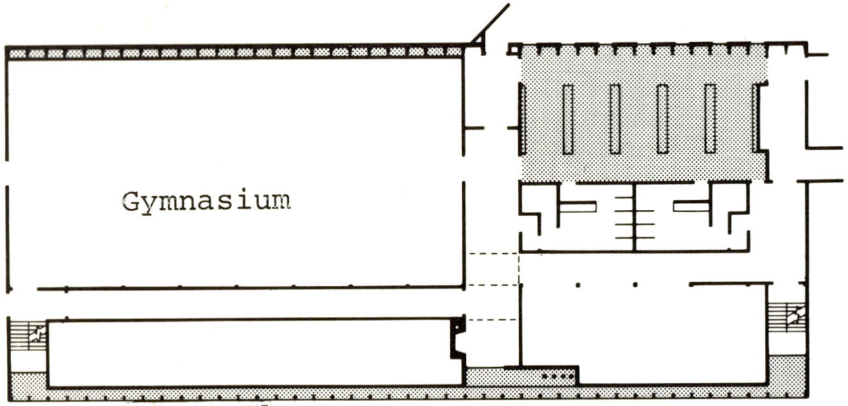

First Floor

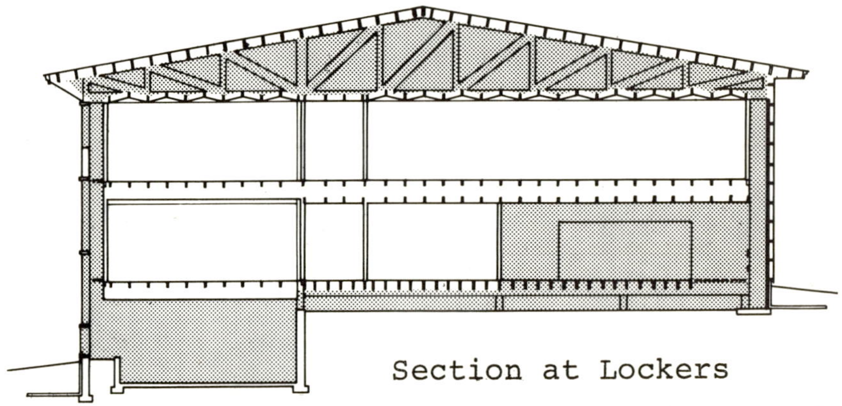

Section at Lockers

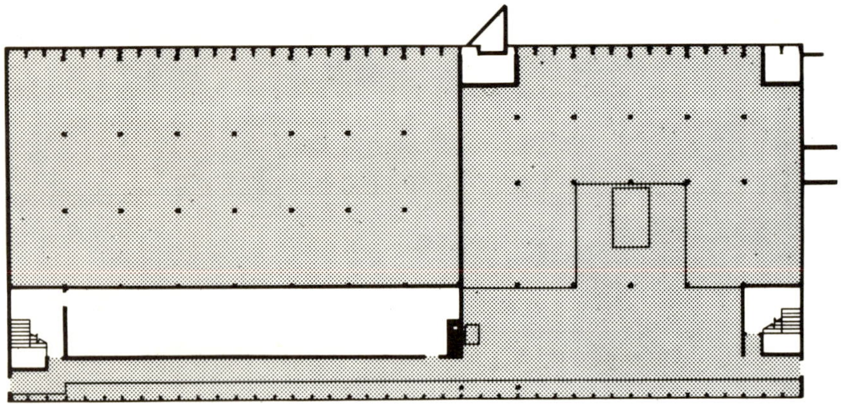

Basement

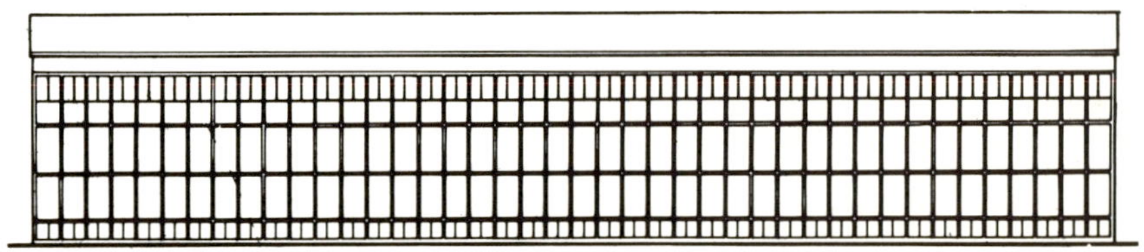

South Elevation

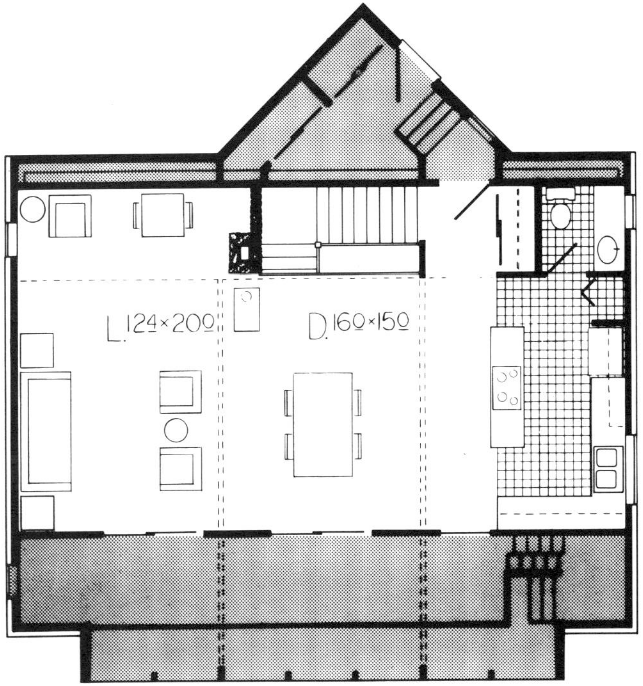
L. 124×200
D. 160×150

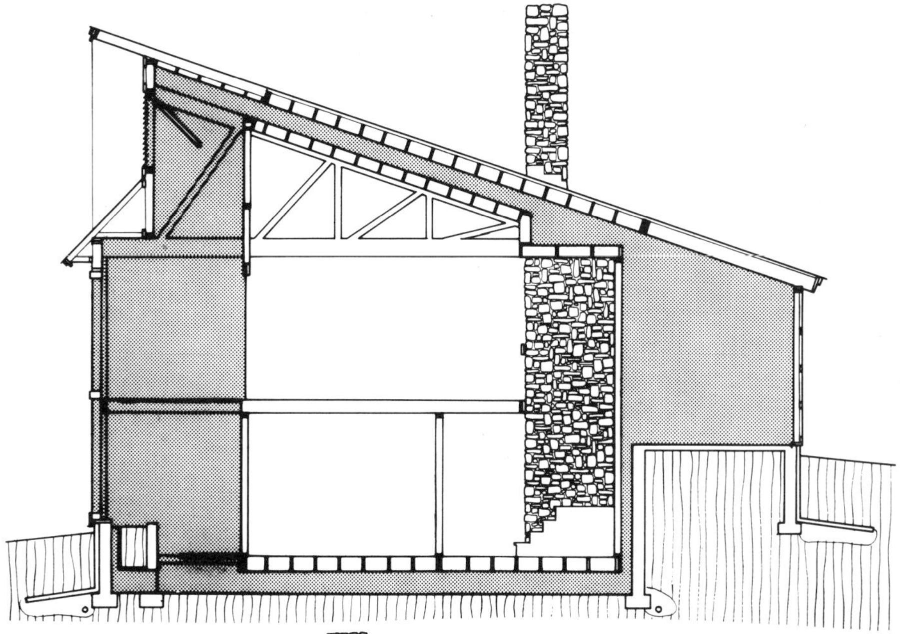

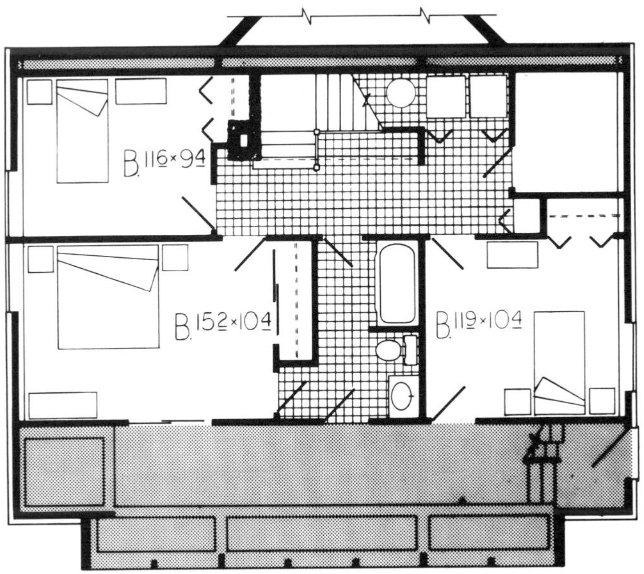
B. 116×94
B. 152×104
B. 119×104

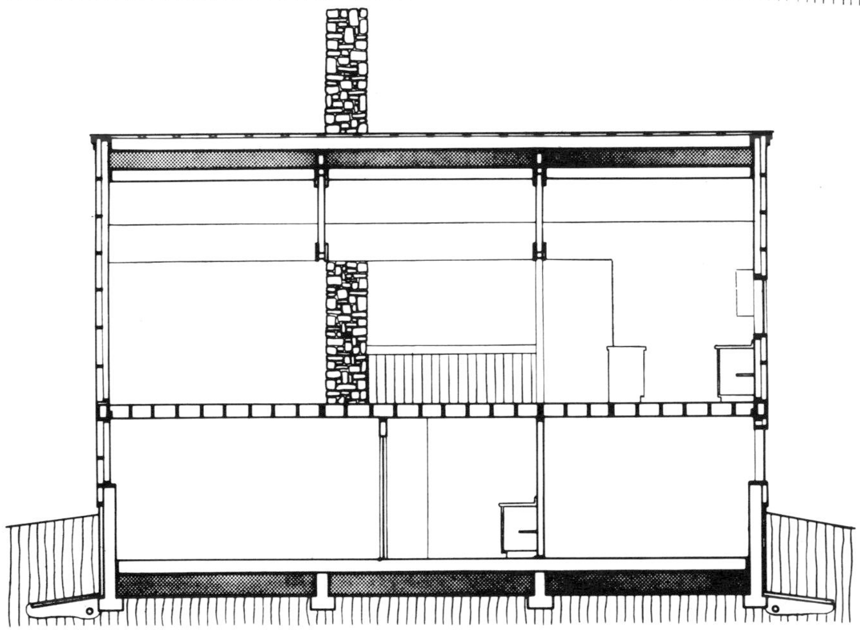

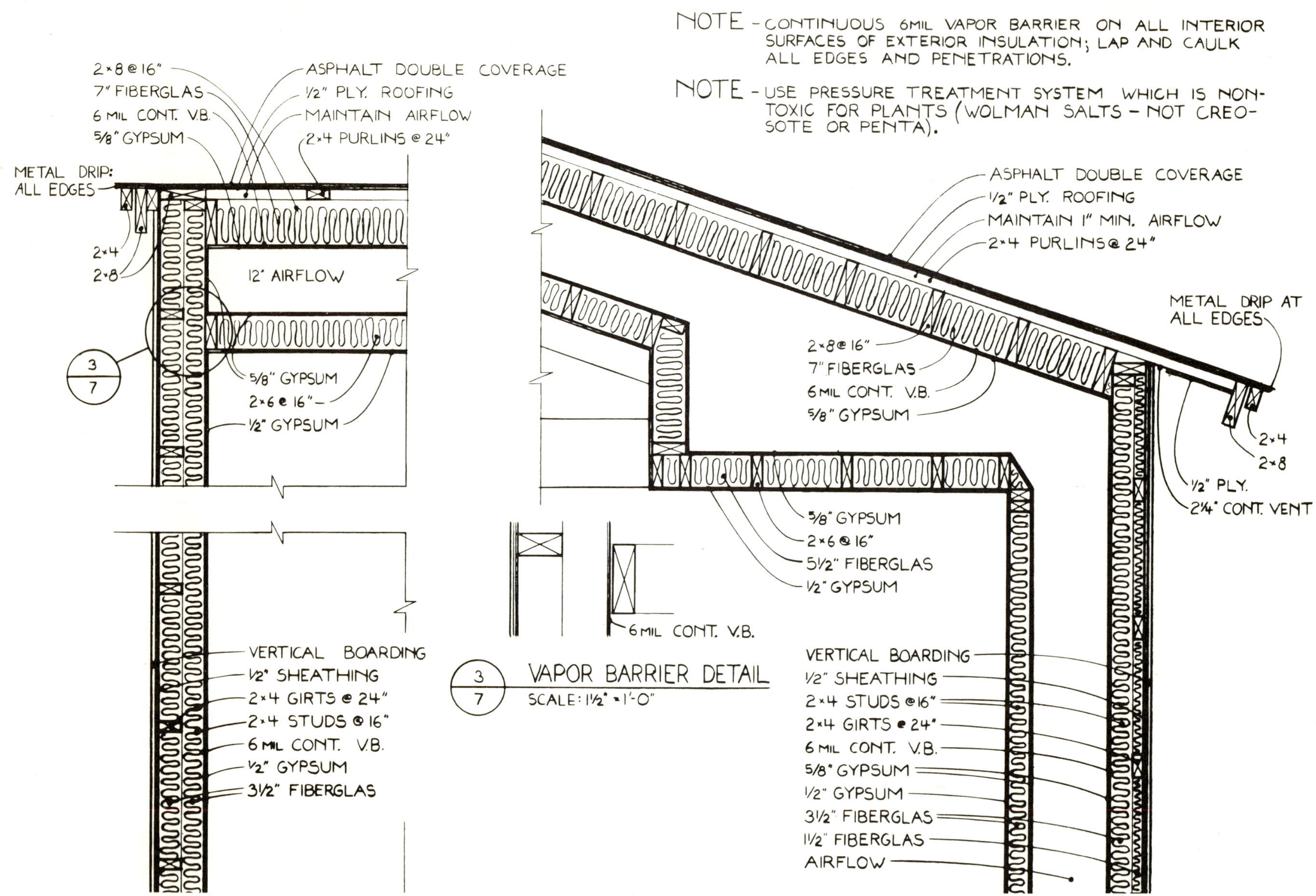
NOTE - CONTINUOUS 6MIL VAPOR BARRIER ON ALL INTERIOR SURFACES OF EXTERIOR INSULATION; LAP AND CAULK ALL EDGES AND PENETRATIONS.
NOTE - USE PRESSURE TREATMENT SYSTEM WHICH IS NON-TOXIC FOR PLANTS (WOLMAN SALTS - NOT CREOSOTE OR PENTA).
2×8 @ 16"
7" FIBERGLAS
6 MIL CONT. V.B.
5/8" GYPSUM
ASPHALT DOUBLE COVERAGE
1/2" PLY. ROOFING
MAINTAIN AIRFLOW
2×4 PURLINS @ 24"
METAL DRIP: ALL EDGES
2×4
2×8
12" AIRFLOW
3/7
5/8" GYPSUM
2×6 @ 16"
1/2" GYPSUM
VERTICAL BOARDING
1/2" SHEATHING
2×4 GIRTS @ 24"
2×4 STUDS @ 16"
6 MIL CONT. V.B.
1/2" GYPSUM
3 1/2" FIBERGLAS
6 MIL CONT. V.B.
3/7 VAPOR BARRIER DETAIL
SCALE: 1 1/2" = 1'-0"
ASPHALT DOUBLE COVERAGE
1/2" PLY. ROOFING
MAINTAIN 1" MIN. AIRFLOW
2×4 PURLINS @ 24"
METAL DRIP AT ALL EDGES
2×8 @ 16"
7" FIBERGLAS
6 MIL CONT. V.B.
5/8" GYPSUM
2×4
2×8
1/2" PLY.
2 1/4" CONT. VENT
5/8" GYPSUM
2×6 @ 16"
5 1/2" FIBERGLAS
1/2" GYPSUM
VERTICAL BOARDING
1/2" SHEATHING
2×4 STUDS @ 16"
2×4 GIRTS @ 24"
6 MIL CONT. V.B.
5/8" GYPSUM
1/2" GYPSUM
3 1/2" FIBERGLAS
1 1/2" FIBERGLAS
AIRFLOW

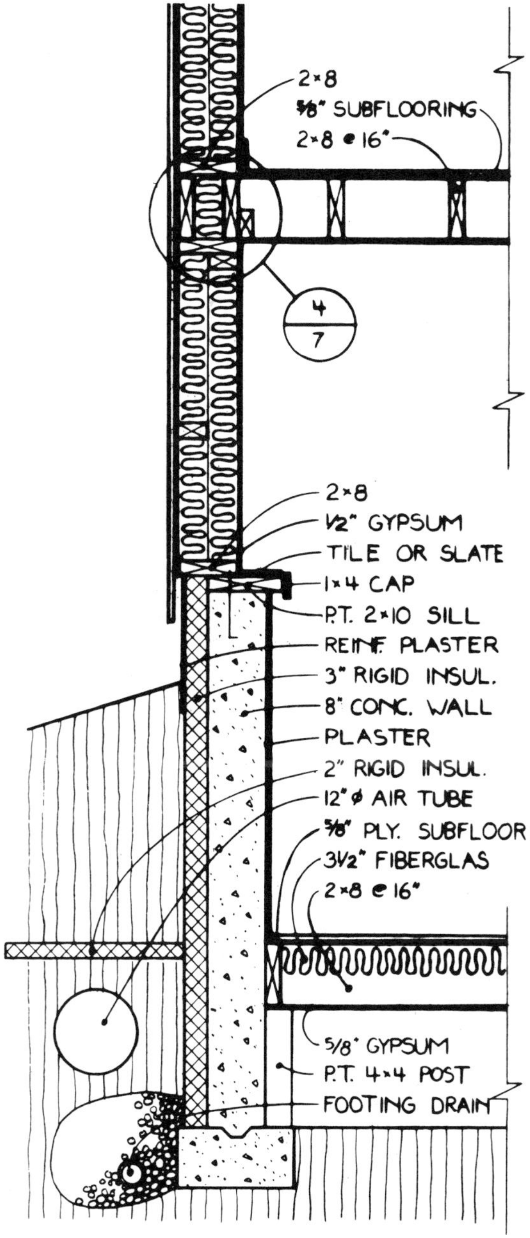

1/7 WEST WALL SECTION

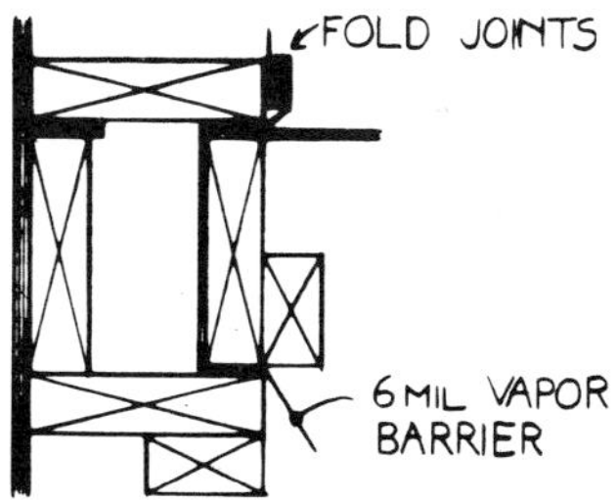

4/7 VAPOR BARRIER DETAIL
SCALE: 1 1/2" = 1'-0"

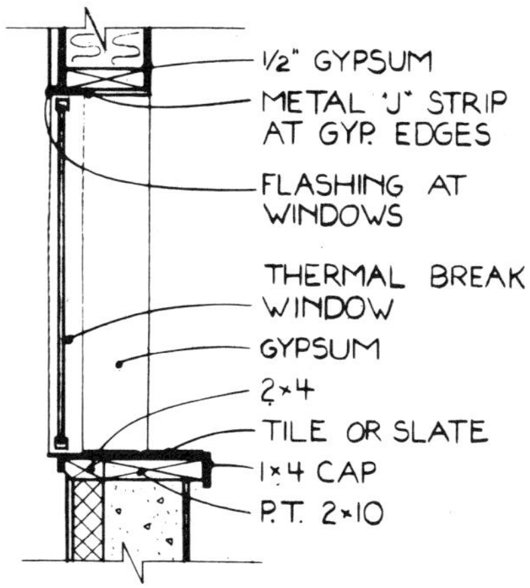

5/7 WINDOW
AT BEDROOMS

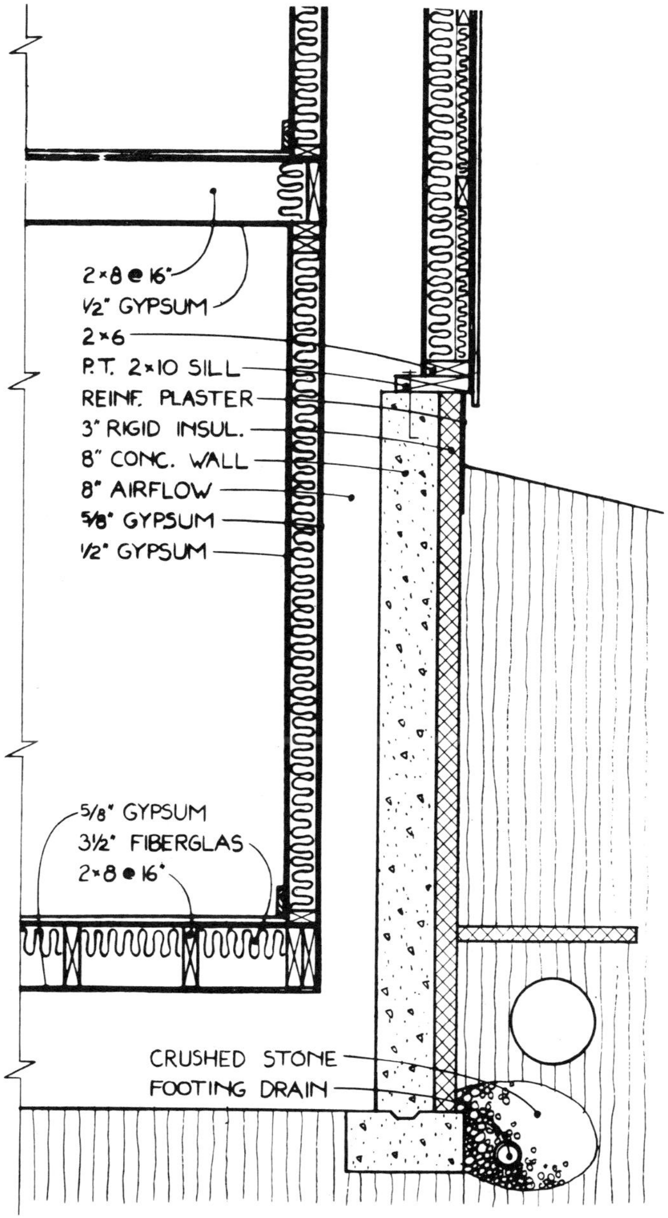

2/7 NORTH WALL SECTION

The sections here are of a different house than that represented by the floor plan. They are included to illustrate a more visible attic connection to the central portion of the sunspace.

The floor plan is another example of our use of framing systems which minimally penetrate the envelope. This house has three framing bays 14' wide.

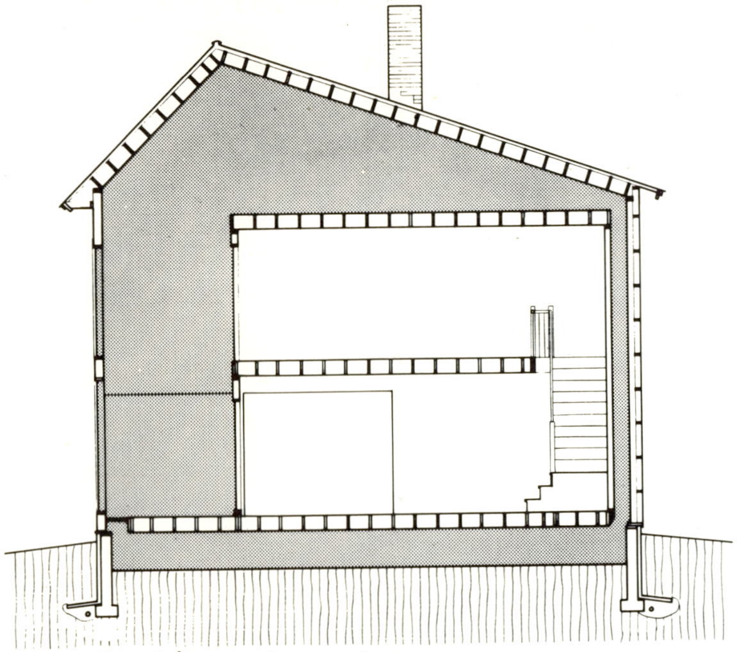

Section at center

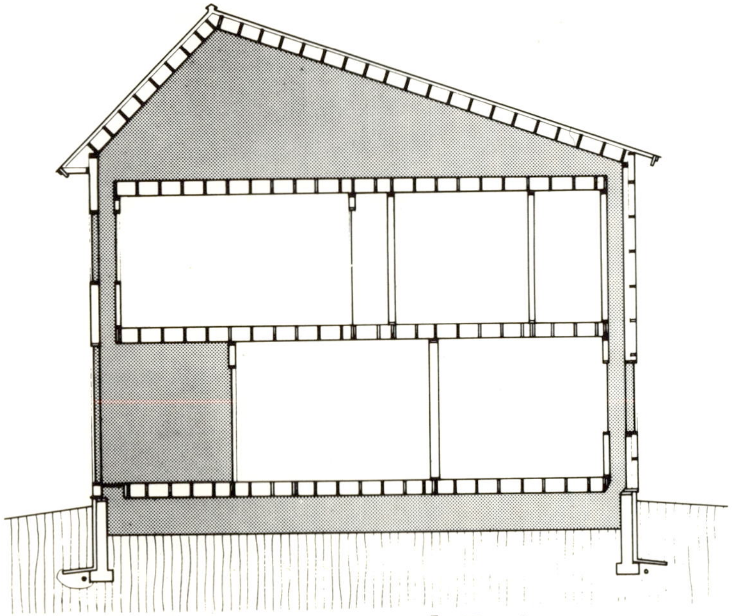

Section at East and West

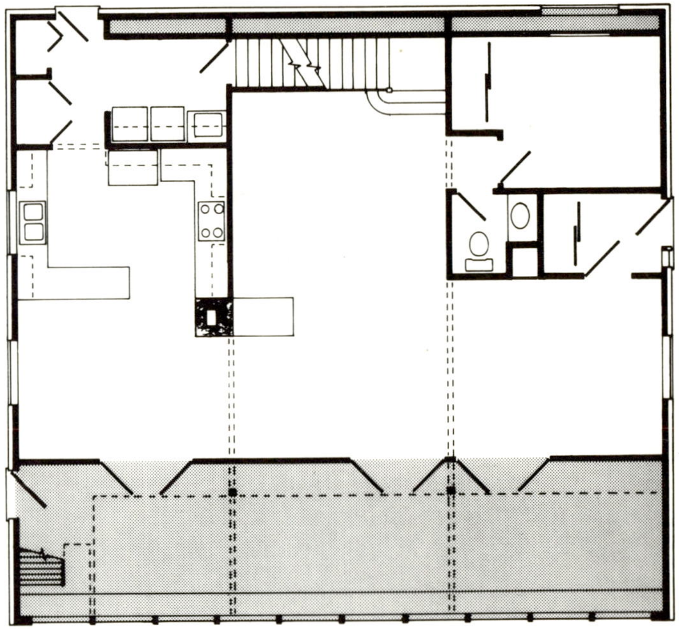

References

Books and Articles

Anderson, Bruce, The Solar Home Book, Brick House Publishing, 3 Main St., Andover, MA 01810 $9.50 (pap.) $13.95 (hard)

ASHRAE Handbook, 1977 Fundamentals, American Society of Heating, Refrigerating and Air-Conditioning Engineers, Inc., 345 East 47th Street, New York, NY 10017

Berk, Jim, Convection Loop Housing, $6, and Design Manual with updates, $25, Solar Clime Designs, Box 9955, Stanford, CA 94305

Butler, Lee Porter, Ekose'a Homes, Ekose'a, 573 Mission St., San Francisco, CA 94105 $24.95

Convection Loops, monthly, "To link up practitioners by exchanging information," $7.50/year. Box AF, Stanford, CA 94305

Earth Sheltered Housing Design, Underground Space Center, University of Minnesota, Pub. by Van Nostrand Reinhold

Labs, Kenneth, "Underground Building Climate," Solar Age, Oct. 1979

McGrath, Ed, How to Build a Superinsulated House, Cold Weather Edition, Project 2020, P.O. Box 81961, College, AK 99708 $4.00

Peoples' Solar Sourcebook, Solar Usage Now, Inc., 450 E. Tiffin St., Bascom, Ohio 44809 $5.00

Shelton, Jay, "On Ducting Outside Air," Solar Age, Sept. 1979

Shurcliff, William A., Superinsulated Houses and Double-Envelope Houses, 19 Appleton Street, Cambridge, MA 02138 $10.00

Smith, Tom and Butler, Lee Porter, The Energy Producing House, Ekose'a, 573 Mission St., San Francisco, CA 94105 $18.95

Other References

Acryl-60, Standard Drywall Products Co., 7800 N.W. 38th St., Miami, Florida

Atlas Industries (insulating panels), 6 Willows Road, Ayer, MA 01432

Double Envelope Computer Program, Norman B. Saunders, 15 Ellis Road, Weston, MA 02193

Dryvit reinforced plaster, Dryvit System Inc., 420 Lincoln Ave., Warwick, RI 02888

HEATPAK energy storage system, Colloidal Materials, Inc., P.O. Box 696, Andover, MA 01810 (617) 475-3276

Kalwall Sunlite, Kalwall Corp., Solar Components Div., P.O. Box 237, Manchester, NH 03105

SINTEF-The Foundation of Scientific and Industrial Research at the University of Trondheim, N 7034 Trondheim-NTH, Norway

SolaRoll (Tube Mat), Bio-Energy Systems, Inc., Box 87, Ellenville, NY 12428

Stereatronics data logger, Norman B. Saunders, 15 Ellis Road, Weston, MA 02193

TEANET, Total Environmental Action, Inc., Harrisville, NH 03450

INDEX

If you're planning to build --

The buffering air envelope is one of many good ways to use the sunshine. For people who want a greenhouse, who want to use very little backup heat or who want to leave the house unoccupied for periods of time with no backup and return to find it comfortable or nearly-comfortable, the envelope can be the best way. It may be the present best way to provide quiet and effortless control of a healthful, humid, plant-refreshed climate.

But it's wise to recognize that we have a lot to learn about using the double shell concepts, and there are other natural sunshine strategies available.

Read what you can find about others' passive solar experience, find the nearest association of solar-conscious people, and gradually tune in yourself to the gently powerful energies generated by the sun and the quietly useful physics of thermal absorption and transfer and storage.

And reach for the most simple ways.

If you have built an air-envelope house, please share your experience with readers of Convection Loops, Box AF, Stanford, CA 94305.

Tom Smith and John Hofacre have prepared a dozen different "Touchstone" plans for envelope houses. Send $1 for information from Positive Technologies, Box 2356, Olympic Valley, CA 95730.

Prints of working plans for Community Builders' houses are shown in reduced size on the next three pages. They are (as are Burns house plans) available at $100/set with extras at $1/sheet (17" x 22").

Each house represents the design constraints of its particular clients as well as our judgements at the time. We try for excellence but our awarenesses are changing, and if we were today preparing these plans for you and your site they could be quite different.

Materials lists are not included.

Community Builders' consulting and design services are available as our time permits. Write for information.

Don Booth is sometimes available to facilitate regional air-envelope workshops. The workshops would draw on the experience of local envelope designers, builders and owners, supplemented with Don's slides and awarenesses.

Community Builders

Shaker Road Canterbury, New Hampshire 03224
Telephone 603/783-4743

Typed by Cindy McCown

ISBN 0-9604422-1-9

Printed by the Sant Bani Press, Tilton, NH

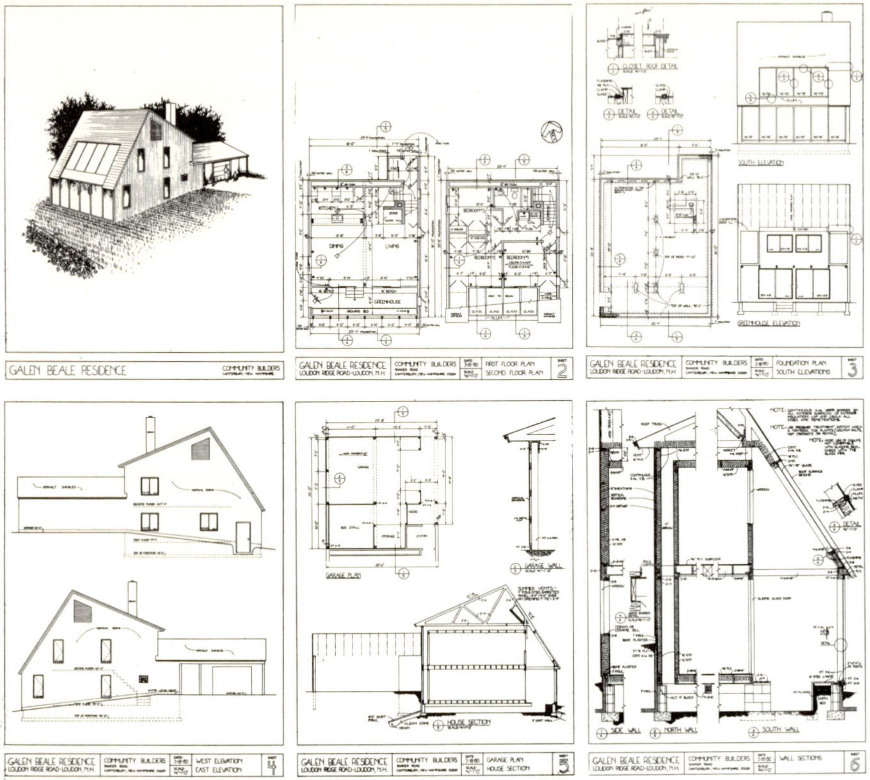
GALEN BEALE RESIDENCE
LOUDON RIDGE ROAD - LOUDON, N.H.
COMMUNITY BUILDERS
FIRST FLOOR PLAN
SECOND FLOOR PLAN
FOUNDATION PLAN
SOUTH ELEVATIONS
WEST ELEVATION
EAST ELEVATION
GARAGE PLAN
HOUSE SECTION
WALL SECTIONS

1

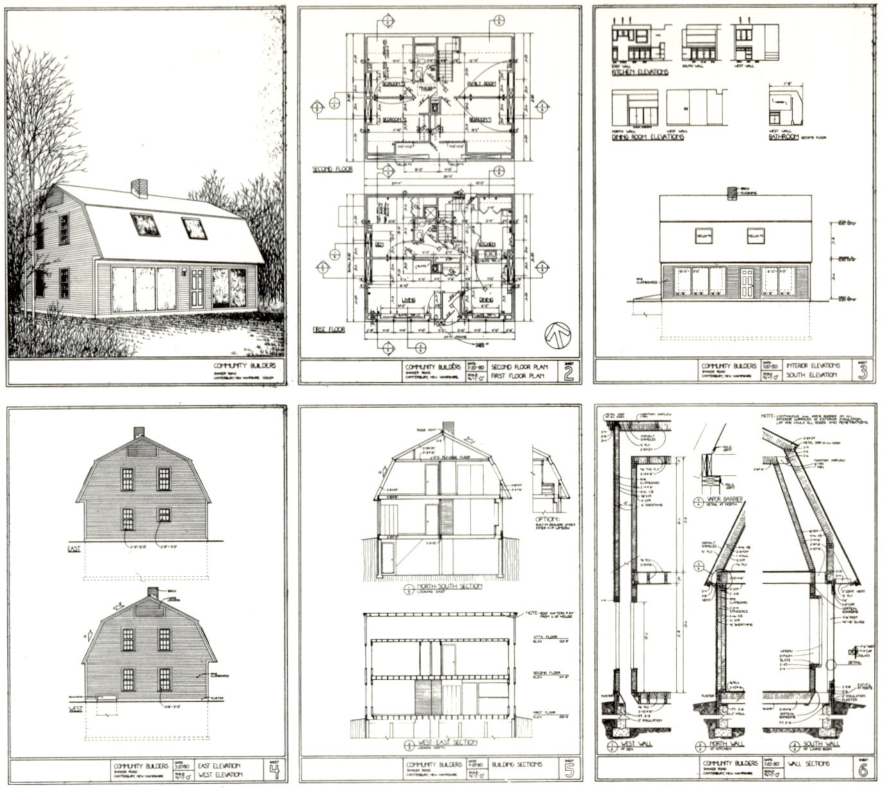
COMMUNITY BUILDERS
SECOND FLOOR PLAN
FIRST FLOOR PLAN
INTERIOR ELEVATIONS
SOUTH ELEVATION
EAST ELEVATION
WEST ELEVATION
BUILDING SECTIONS
WALL SECTIONS

2

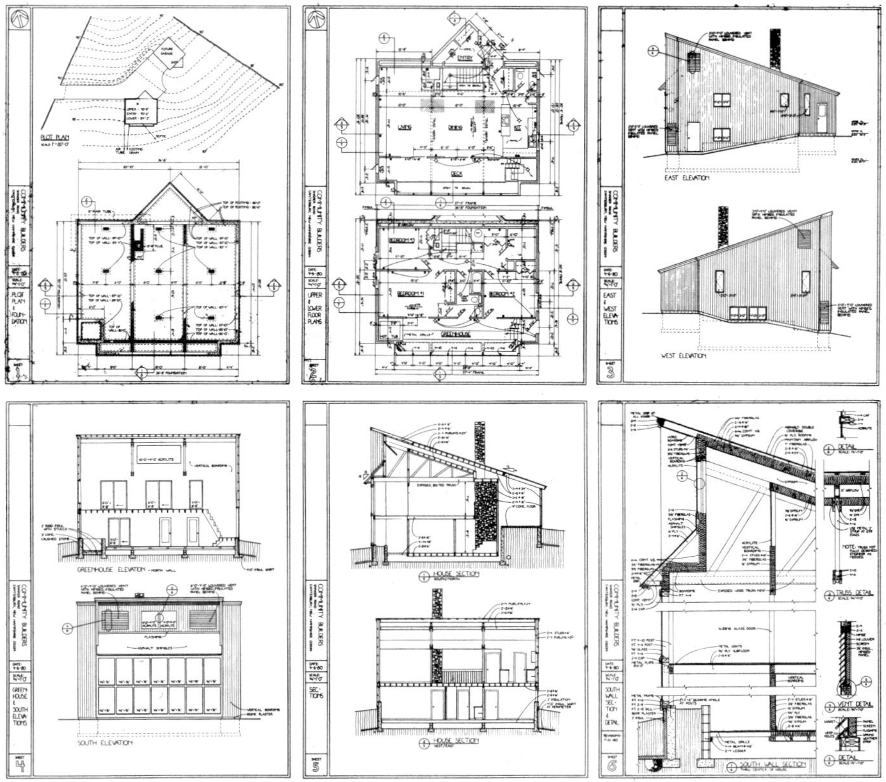
PLOT PLAN
COMMUNITY BUILDERS
EAST ELEVATION
WEST ELEVATION
GREENHOUSE ELEVATION
SOUTH ELEVATION
HOUSE SECTION
SOUTH WALL SECTION

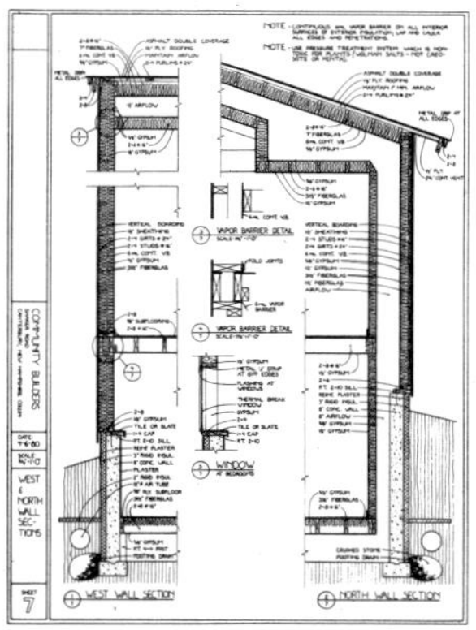
COMMUNITY BUILDERS
WEST WALL SECTION
NORTH WALL SECTION

3

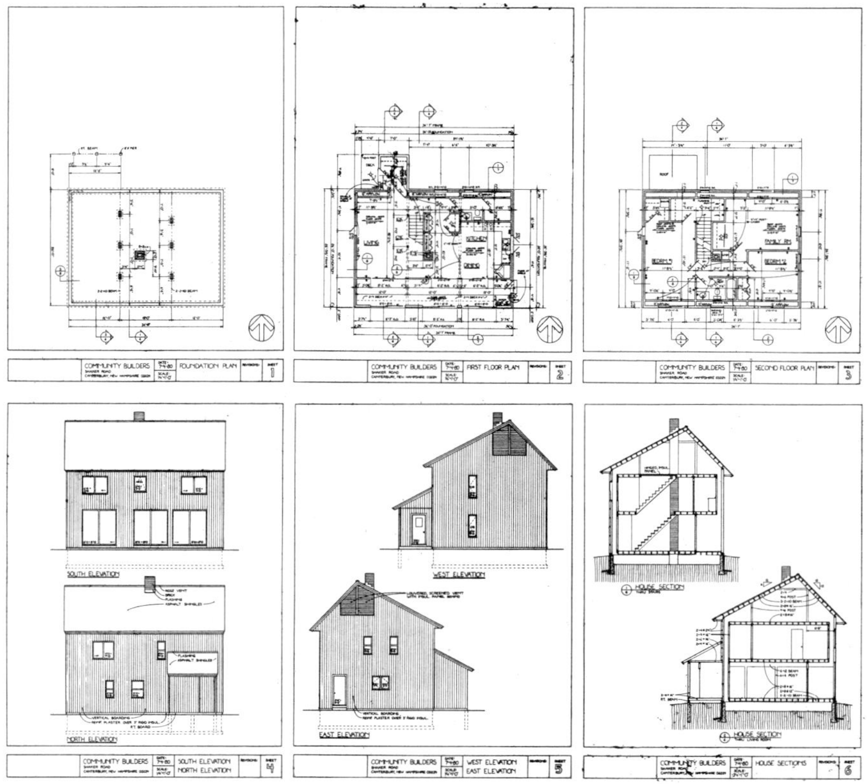
COMMUNITY BUILDERS
FOUNDATION PLAN
FIRST FLOOR PLAN
SECOND FLOOR PLAN
SOUTH ELEVATION
NORTH ELEVATION
WEST ELEVATION
EAST ELEVATION
HOUSE SECTIONS

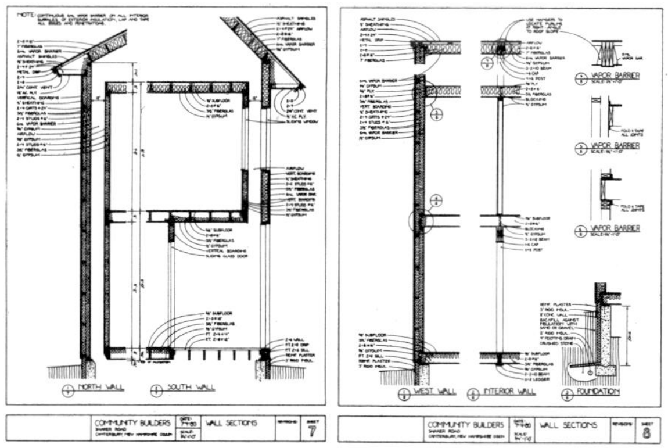
COMMUNITY BUILDERS
WALL SECTIONS
NORTH WALL
SOUTH WALL
WEST WALL
INTERIOR WALL
FOUNDATION

4

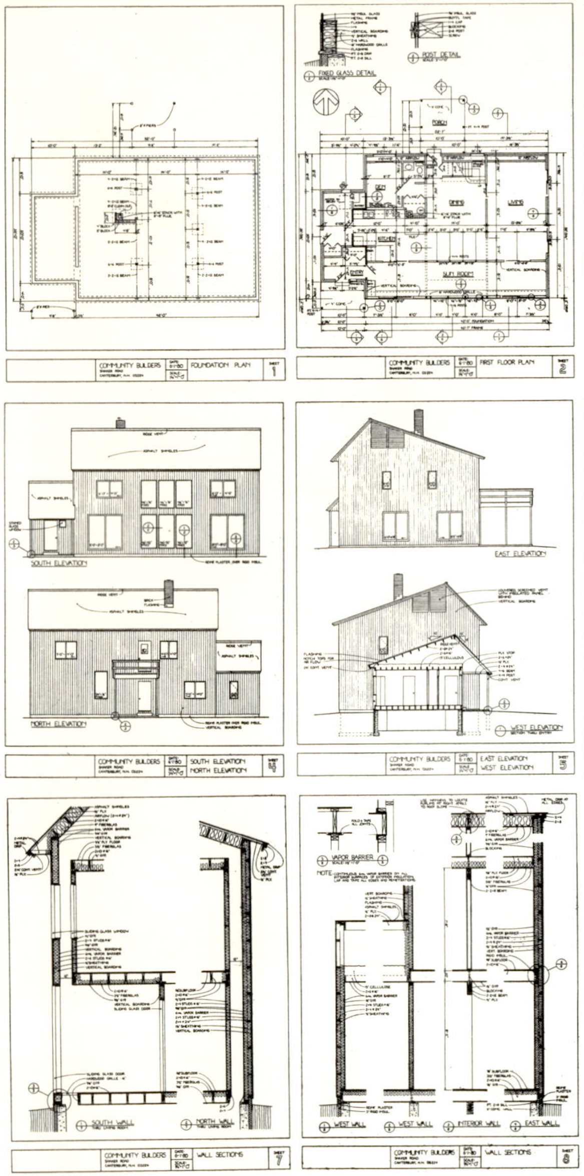
COMMUNITY BUILDERS
FOUNDATION PLAN
FIRST FLOOR PLAN
SECOND FLOOR PLAN
SOUTH ELEVATION
NORTH ELEVATION
EAST ELEVATION
WEST ELEVATION
HOUSE SECTIONS
WEST SECTION
CENTER SECTION
WALL SECTIONS
SOUTH WALL
NORTH WALL
WEST WALL
INTERIOR WALL
EAST WALL

5

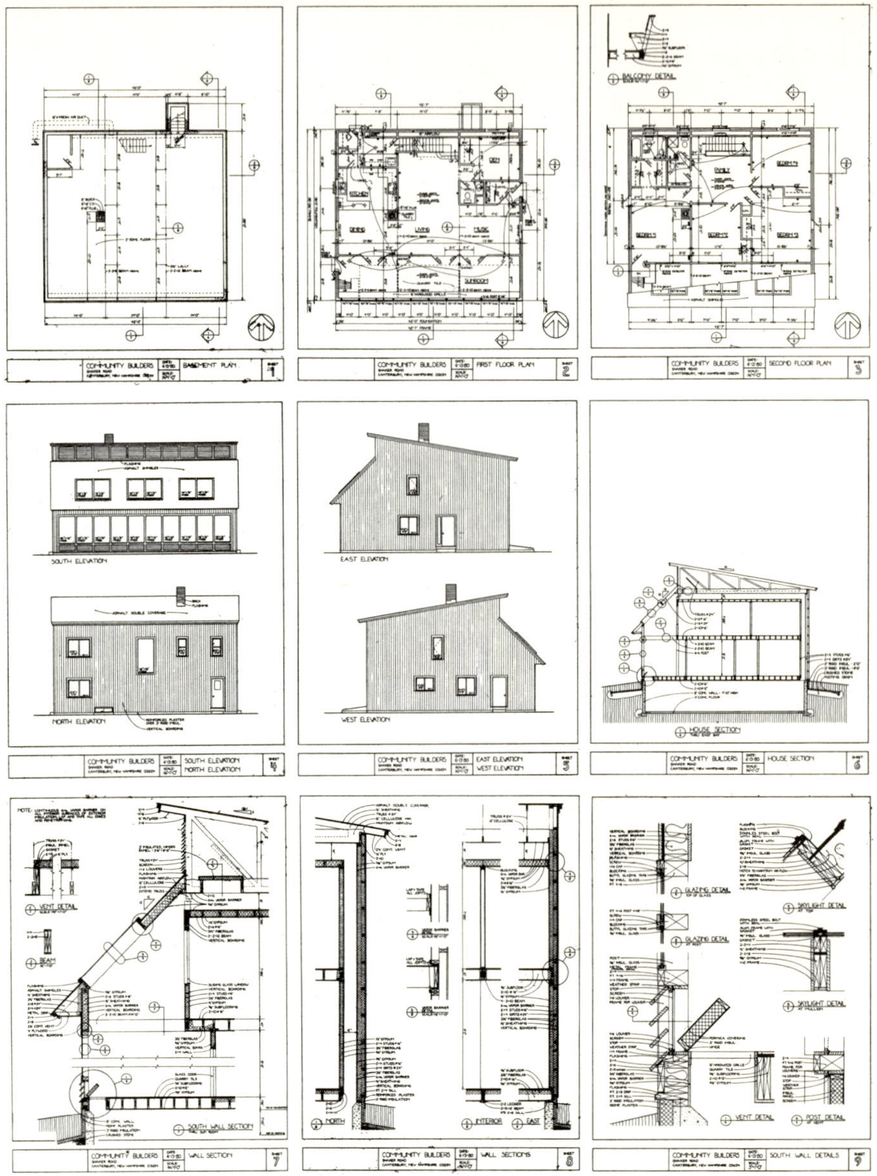
COMMUNITY BUILDERS
BASEMENT PLAN
FIRST FLOOR PLAN
SECOND FLOOR PLAN
BALCONY DETAIL
SOUTH ELEVATION
NORTH ELEVATION
EAST ELEVATION
WEST ELEVATION
HOUSE SECTION
WALL SECTION
SOUTH WALL SECTION
WALL SECTIONS
NORTH
INTERIOR
EAST
SOUTH WALL DETAILS
GLAZING DETAIL
SKYLIGHT DETAIL
VENT DETAIL
POST DETAIL

6

273
21513